COSMOS, BIOS, THEOS

· · · · · · · · · ·

COSMOS, BIOS, THEOS

Scientists Reflect on Science, God,
and the Origins of the Universe,
Life, and *Homo sapiens*

EDITED BY

Henry Margenau

Roy Abraham Varghese

Open Court
Chicago and La Salle, Illinois

Open Court Publishing Company is a division of
Carus Publishing Company.

© 1992 by Open Court Publishing Company

First printing 1992
Second printing 1993
Third printing 1994

Printed and bound in the United States of America.

Library of Congress Cataloging-in-Publication Data

Cosmos, bios, theos : scientists reflect on science,
God, and the origins of the universe, life, and
homo sapiens / edited by Henry Margenau,
Roy Abraham Varghese.
        p.   cm.
    Includes bibliographical references and index.
    ISBN 0-8126-9185-7. — ISBN 0-8126-9186-5 (pbk.)
    1. Science —Philosophy.   2. Religion and science —
1946–   3. Scientists—Attitudes.   I. Margenau,
Henry, 1901–   II. Varghese, Roy Abraham.
Q175.3.C68 1992
215—dc20   92-7685
                                    CIP

This book has been reproduced in a print-on-demand
format from the 1994 Open Court printing.

**To order books from Open Court,
call toll-free 1-800-815-2280.**

CONTENTS

· · · · · · · · · ·

PART ONE
· · · · · · · · · ·

Astronomers, Mathematicians, and Physicists   27

PART TWO
. . . .
Biologists and Chemists   *137*

PART THREE

· · · · · · · · · ·

The Existence of God and the Origin of the Universe: A Debate between an Atheist and a Theist   *225*

PREFACE

· · · · · · · · · ·

Cosmos, Bios, Theos is a portfolio of perspectives on the relationship between the scientific enterprise and the religious view of reality. Contributors include over twenty Nobel Prize winners and distinguished scientists from different disciplines. In this anthology, they reflect on the origins of the universe, life, and *Homo sapiens*, on science and religion, and on the existence of God.

Cosmos, Bios, Theos makes no pretension to being a statistically significant survey of the religious beliefs of modern scientists. The scientists interviewed for this anthology are, for the most part, known to be theistic or at least sympathetic to a religious view of reality. For this reason, it must be clearly understood that the book does not purport to show that most or even many scientists are theists. In point of fact, there are many modern scientists who are atheists actively opposed to any form of religion. Moreover, a number of the contributors to this anthology make it clear that they are simply uninterested in religion.

Notwithstanding these caveats, the contributions in this collection are significant in their own right. In the first place, the idea that an eminent scientist would affirm the existence of God on rational grounds does at least pique one's curiosity—especially in view of the popular assumption that religious belief is an anachronism or an aberration in the Age of Science. Secondly, this inquiry into the interface between science and religion is a continuation of a quest begun by some of the great pioneeers of modern science. Co-editor Henry Margenau notes that "several modern scientists and scientific theories have been surprisingly sympathetic to religious issues. I recall that my late teachers/colleagues/friends, Einstein, Schrödinger, and Heisenberg, who were all distinguished scientists, had a passionate interest in religious questions". Finally, the metascientific implications of recent developments in science have inspired a spurt of popular quasi-theological works by contemporary scientists. Examples include Stephen Hawking's *A Brief History of Time,* Paul

Davies's *God and the New Physics* and *The Cosmic Blueprint,* John Leslie's *Universes,* Robert Jastrow's *God and the Astronomers,* and several others. The issues addressed in these works certainly deserve further exploration.

All but a few of the contributions to *Cosmos, Bios, Theos* are responses to the six questions below:

1 What do you think should be the relationship between religion and science?

2 What is your view on the origin of the universe: both on a scientific and—if you see the need—on a metaphysical level?

3 What is your view on the origin of life: both on a scientific level and—if you see the need—on a metaphysical level?

4 What is your view on the origin of *Homo sapiens*?

5 How should science—and the scientist—approach origin questions, specifically the origin of the universe and the origin of life?

6 Many prominent scientists—including Darwin, Einstein, and Planck—have considered the concept of God very seriously. What are your thoughts on the concept of God and on the existence of God?

The pieces by Sir John Eccles, Professor Robert Jastrow, and Professor Brian Josephson are transcripts of interviews they had with co-editor Roy Abraham Varghese. Professors Clifford Matthews, Sir Nevill Mott, Arno Penzias, Abdus Salam, and George Wald submitted full length essays in response to the six questions.

While an anthology of this nature can hardly hope to give "equal time" for all contending points of view, it does contain a spirited debate on the existence of God between the well-known atheist Antony Flew and prominent philosopher of religon H. D. Lewis. This debate is important in defining and clarifying the origin questions as they relate to the existence of God. The debate between Flew and Lewis is followed by an essay on origins in science and religion by Professor William Stoeger. The anthology ends with a paper by Professor Eugene Wigner on relativity, quantum theory, and the mystery of life.

INTRODUCTION

Roy Abraham Varghese

I have never found a better expression than "religious" for this trust in the rational nature of reality and of its peculiar accessibility to the human mind. Where this trust is lacking science degenerates into an uninspired procedure. Let the devil care if the priests make capital out of this. There is no remedy for that.

> —*Albert Einstein*[1]

Certain it is that a conviction, akin to religious feeling, of the rationality or intelligibility of the world lies behind all scientific work of a higher order. . . . This firm belief, a belief bound up with deep feeling, in a superior mind that reveals itself in the world of experience, represents my conception of God.

> —*Albert Einstein*[2]

There can never be any real opposition between religion and science; for the one is the complement of the other.

> —*Max Planck*[3]

In the history of science, ever since the famous trial of Galileo, it has repeatedly been claimed that scientific truth cannot be reconciled with the religious interpretation of the world. Although I am now convinced that scientific truth is unassailable in its own field, I have never found it possible to dismiss the content of religious thinking as simply part of an outmoded phase in the consciousness of mankind, a part we shall have to give up from now on. Thus in the course of my life I have repeatedly been compelled to ponder on the relationship of these two regions of thought, for I have never been able to doubt the reality of that to which they point.

> —*Werner Heisenberg*[4]

Up to now, most scientists have been too occupied with the development of new theories that describe what the universe is to ask the question why. . . . If we find the answer to that, it would be the ultimate triumph of human reason—for then we would know the mind of God.

> —*Stephen Hawking*[5]

Stranger than the strangest concepts and theories of science is the appearance of God on the intellectual horizon of late twentieth-century science. Pioneers and giants of modern science like Einstein, Planck, and Heisenberg were equally at home with the "hard facts" of science and with a theological frame of reference (as evidenced by their statements above). The paradox persists to this very day. Einstein once said "I want to know how God created this world. . . . I want to know His thoughts, the rest are details"[6]; Stephen Hawking, the theoretical physicist who is often described as Einstein's successor, has declared in his recent bestseller, *A Brief History of Time*, that our goal should be to "know the mind of God".[7] And astrophysicist Robert Jastrow begins *God and the Astronomers*, his celebrated survey of modern cosmology, with a remarkable observation: "For the scientist who has lived by his faith in the power of reason, the story ends like a bad dream. He has scaled the mountain of ignorance; he is about to conquer the highest peak; as he pulls himself over the final rock, he is greeted by a band of theologians who have been sitting there for centuries".[8]

. . . .
The Scientific Enterprise as a Religious View of Reality

The scientific enterprise is fundamentally a quest for an explanation of "what is" in terms of relations and laws, principles and patterns. "Explanation" here is understood as the "how" and the "why" of the "what", where the "why" is not necessarily a question of purpose or goal but refers also to inquiries about the *cause* of a state of affairs. "The aim of scientific research is not only to discover and describe events and phenomena in the world but also, and more importantly, to explain why these events and phenomena occur as they do".[9]

This quest for an explanation of "what is" presupposes not only that there is an explanation (or explanations) for the objects of scientific inquiry but that (a) reality is intelligible, accessible to the human mind, and (b) reality is rational, conforming to at least some a priori conceptions of rationality.

The significance of the intelligibility and rationality of reality must be further explicated. If science is to "work", the universe must be intelligible in the sense that we can, in principle, come to know it. And, as Professor Hugo Meynell notes, "We must be in principle capable of recognising the truth before we reach it, if we are ever to recognise it as such when we do so. Could scientists ever have achieved the knowledge of the world which they now have, except on this assumption, which has very seldom been *effectively* doubted—the qualification is important—that the framing of hypotheses and the testing of them against the evidence of experience is the correct method of discovering what is so about the world? And does not the fact that we know *a priori* how to find out about things imply, short of idealism, a certain corresponding nature and structure in things themselves?"[10]

About the rationality presupposed and revealed by science, Einstein said, "Whoever has undergone the intense experience of successful advances in this domain (science) is moved by profound reverence for the rationality made manifest in existence . . . the grandeur of reason incarnate in existence".[11]

Neither intelligibility nor rationality are presuppositions in the sense of hypotheses that are verified by continuing scientific development, a point made with characteristic clarity by Frederick Copleston, the great historian of philosophy:

> The intelligibility of being is not a presupposition in the sense of hypothesis. If it is called a presupposition, what is meant, or ought to be meant, is that it is a condition of knowledge. It is not a condition of which the mind must be aware before it knows anything: it is simply a condition which must be fulfilled if we are to know anything. That it is fulfilled is apprehended implicitly in the act of knowing anything. Explicit apprehension is due to subsequent reflection.[12]

In the constant quest for explanations that ground the perception of intelligibility and rationality, the human mind confronts the challenge of explaining the existence of the universe. "Why is there something rather than nothing?" asked Leibniz[13]—a "why" question that is just as much about ultimate origin as it is about ultimate purpose. The question is equally forceful whether the universe has trillions of galaxies or whether all that exists is a grain of sand. How did it—the galaxies or the grain of sand—get here? It is now in being. How and why was it brought into being? Did it create itself? Or was it always here with no beginning and, presumably, no end? But to say that it always existed is still not an *explanation* of its existence. It is, rather, a putative, highly speculative description that is, by the nature of the case, incapable of scientific verification (how is it possible to *prove* that matter and energy have no beginning and no end?). And a description is not an explanation. The assertion that the universe always existed is not, to my mind, an explanation for this reason: even if we admit the assumption of an eternally existing universe we are still left with the problem of explaining and accounting for the phenomenon of an eternally existing universe.

Science and scientists have not ignored the puzzle, and many of the spectacular discoveries and theories of modern science are by-products of the quest for an explanation of the existence of the universe. By-products they may be, but they are clearly not end-products. The mathematics and the mechanisms behind the processes that culminated in the universe we inhabit have been the objects of plausible and often fruitful speculation. But the question of ultimate origin—an ultimate explanation for the mathematics and the mechanisms—continues to elude and baffle the most ingenious theorists. A review of influential contemporary cosmological models highlights the depth of the mystery.

The idea that matter and the universe are infinite in space and time has been taken for granted for centuries by several schools of thought. The postulate that matter is eternal cannot, of course, be verified scientifically or proved philosophically. To be tested scientifically, this hypothesis would require an experiment of eternal duration (and it would have to assume without proof that matter and energy are indestructible). And philosophically the hypothesis begs the bottom-line question: *how* did matter and the universe come into being? Such a question can be addressed even with regard to an eternally existing universe, because we want to know how it is that there *is* a universe with the property of eternal existence. It is not enough to say *that* the universe was always here and that we should not ask how it got here or how, if it is eternal, the phenomenon of an eternally existing universe can be explained. Why should we not ask these questions? In any case, the notion that the universe has infinite mass—a notion that, in its Euclidean form, was conclusively exorcised from science by the General Theory of Relativity—bristles with bizarre scientific consequences.

Einstein's General Theory of Relativity was a leap of genius that all but created a universe—and with it modern scientific cosmology—out of mere "matter in motion". The inspired conceptions of gravity as geometry, of matter "curving" space, of inferring both the total mass of the universe and the resultant radius of curvature from the average density of matter and from the gravitational constant, have all changed the course of modern science. This new understanding, in Hawking's words, "was to revolutionize our view of the universe. The old idea of an essentially unchanging universe that could have existed, and could continue to exist, forever was replaced by the notion of a dynamic, expanding universe that seemed to have begun a finite time ago, and that might end at a finite time in the future".[14] Professor Stanley Jaki points out also that Einstein's "General Relativity, the first consistently scientific treatment of the universe as the totality of gravitationally interacting entities, reassured him in his previous instinctive conviction that the universe was real and fully rational".[15]

One cosmological implication of General Relativity—an implication that Einstein himself found disconcerting—was the idea that the universe could not be static and must be either expanding or contracting. By the late 1920s experimental and theoretical evidence clearly pointed to an expanding universe. Today, as a result of the equations of General Relativity, observational studies of galaxies receding from each other, the measurement of the background radiation that fills space, and a host of other phenomena, there is general acceptance of a Big Bang model of the origin of the universe. "Most scientists", writes Heinz Pagels, "maintain that the universe evolved from a hot, dense gas of quantum particles which subsequently expanded rapidly—an explosion called the 'hot big bang'. Everything in the universe is a remnant of that explosion".[16] (The currently influential cosmological theories on the origin of the universe are usefully surveyed in the *Scientific American* report on the June 1990 Nobel symposium on "the birth and early evolution of our universe".[17])

The thesis that the universe had a beginning of some sort lends both urgency and plausibility to the quest for an ultimate explanation for the existence of the universe. Dr. Arno Penzias, one of the pioneer Big Bang experimentalists, focuses on this factor in his essay in this volume: "Today's dogma holds that matter is eternal. The dogma comes from the intuitive belief of people (including the majority of physicists) who don't want to accept the observational evidence that the universe was created—despite the fact that the creation of the universe is supported by all the observable data astronomy has produced so far. As a result, the people who reject the data can arguably be described as having a 'religious' belief that matter must be eternal. . . . If the universe hadn't always existed, science would be confronted by the need for an explanation of its existence" (page 79f.).

That there has to be an explanation is evident to the scientific mind. But it is equally evident that this explanation cannot be verified or falsified with scientific tools if it involves absolute nothingness at any point—as explained later in this anthology by Professor William Stoeger. (See Part Four.)

Theoreticians at the cutting-edge of cosmology have tried to overcome the obstacles to explanations of the Big Bang in scientific terms by probing the ultimate and the fundamental laws and mechanisms behind the processes detected by the experimentalists. The most influential approaches to this level of explanation include the Oscillating Universe hypothesis and the Quantum Gravity, Vacuum Fluctuation, and Inflationary Universe models.

According to the Oscillating Universe hypothesis, the expansion of the universe will cease some time in the future and the universe will contract to its initial state. This contraction will be followed by another explosion, then a contraction, then an explosion, and so on without end (or beginning). Attractive though it is in some respects, this hypothesis merely postpones or evades the questions of ultimate origin and ultimate explanation indefinitely. How is the existence of this whole mechanism of eternal expansion and contraction to be explained? On a scientific level, the hypothesis seems incapable of verification or falsification and lacks the kind of experimental evidence that grounds the Big Bang model. Professor Jastrow points out that the evidence is against the Oscillating Universe hypothesis "because the latest word is that there is not enough matter in the Universe to halt the expansion and bring the Universe together again. Furthermore each time the Universe collapses, it melts down the world that existed, and if it does emerge from the collapse, an entirely new world results. This is so far beyond our experience in religion or in science, that it seems to me it is not worth talking about" (page 46). Be this as it may, the most important consideration is the question of ultimate origin. And here the Oscillating Universe hypothesis cannot explain how a universe with oscillating tendencies came to be—or how it continues to exist.

By far the most fascinating theory of the origin of the universe is the Quantum Gravity model. The Big Bang theory, in its standard formulations, suggests that there are singularities in the universe, points at which the laws of physics break down. The notion of singularities has not been attractive even to Stephen Hawking, who showed that General Relativity indicates that the universe began in a singularity. In his recent works, Hawking has sought to eliminate the necessity for a singularity at the origin of the universe by explaining the Big Bang in terms of a theory of quantum gravitation. This remarkable theory is an attempt to unite the science of the Macroverse (General Relativity) with the science of the Microverse (quantum physics) so that singularities cease to exist. By applying the principles of quantum physics to the Big Bang itself, and by developing a complex conception of "imaginary time", Hawking has tried to show that space-time is finite but unbounded and that the laws of science do not break down at any point. In Hawking's words: "When we combine quantum mechanics with general relativity, there seems to be a new possibility that did not arise before: that space and time together might form a finite, four-dimensional space without singularities or boundaries, like the surface of the earth but with more dimensions".[18] Time, it appears, is literally "spaced out", and the boundary condition of this four-dimensional space is that it has no boundary.

Although Hawking points out that his theory is only a proposal—"So far, my work on how the no-boundary proposal determines how the universe comes out of the big bang is still in a preliminary stage. . . . I don't think you can completely test it by observation"[19]—and that it cannot be deduced from some other principle, it is clear that verification of its validity will not resolve the question at issue here. Hawking himself asks at the end of *A Brief History of Time*: "What is it that breathes fire into the equations and makes a universe for them to describe? The usual approach of science of constructing a mathematical model cannot answer the questions of why there should be a universe for the model to describe. Why does the universe go to all the bother of existing?".[20] Hawking's colleague, Roger Penrose, remarks: "Once you have put more and more of your physical world into a mathematical structure, you realize how profound and mysterious this mathematical structure is. How you can get all these things out of it is very mysterious".[21]

Moreover, as pointed out by a reviewer in *Nature*, Hawking's speculations center on "what he considers to be the most common view of God's activity—that God started off the Universe and then let it evolve without intervention". Traditional theism, however, holds:

that God creates and sustains the entire Universe rather than just the beginning. Whether or not the Universe has a beginning has no relevance to the question of its creation, just as whether an artist's line has a beginning and an end, or instead forms a circle with no end, has no relevance to the question of its being drawn. Of course, if the no-boundary proposal is correct, it does have interesting implications for the beauty, elegance and simplicity of the Universe God did create. . . . Choosing the no-boundary state and then actually carrying out the immense task of the creation of the Universe in this state is a far cry from Carl Sagan's claim in his introduction to Hawking's book of 'nothing for a Creator to do'".[22]

Stoeger's comments on Hawking's statements reiterate the nature of the problem: "Any physical model of the universe—even though it be the definitive unified field theory—will apparently not be able to account intelligibly for the existence of the universe by itself. It will need some underlying existential ground or support, which must be presupposed by science and which science itself, as it is presently conceived and practiced, cannot completely reveal" (page 258).

Vacuum Fluctuation theories of the origin of the universe, in combination with inflationary views of the expansion of the universe, depict the Big Bang as the end-product of the quantum fluctuation of a primordial vacuum. A vacuum is seen here as being saturated with quantum fields and as being subject to fundamental fluctuations (a prediction of quantum physics). When strong enough, the fluctuating energy fields appear briefly as "virtual" particles and then disappear. Professor Timothy Ferris outlines the route from vacuum to universe:

> According to the inflationary scenario, the radius of the universe increased by some 10^{50} times, from smaller than a proton to larger than a softball, during the first 10^{-30} second of time. During this brief but critical period the universe was but a vacuum. Its potential mass and energy could not yet manifest itself as particles, because space was expanding too fast for the particles to congeal out of the vacuum. Technically, one described this condition by saying that the vacuum was hung up in a symmetrical state during a phase transition. . . . The cosmic vacuum remains empty even after falling below the temperature at which particle production ordinarily would take place. Indeed, it is this hang-up that drives the expansion: The latent energy is tied up in what is called a zero-value Higgs field, and the field acts as an

engine that inflates the dimensions of cosmic space, driving the expansion so that the empty universe balloons in perfect Platonic sphericity. Eventually, (meaning after about 10^{-30} second) the quantum instability of the situation catches up with it, and the expansion abruptly slows to a linear rate. When that happens the energy latent in the vacuum precipitates out as particles and antiparticles. The particles mutually annihilate, and the resulting flood of energy inaugurates the big bang.[23]

This conception of the origin of the universe—an extension of the standard Big Bang model—has inspired the idea that our universe is just one bubble in a sea of multiple bubble-universes generated by other quantum fluctuations. Professor Sidney Coleman of Harvard University speculates that the various universes are connected via fluctuations in the space-time field called quantum wormholes.

Like the Quantum Gravity model, Vacuum Fluctuation and Inflationary Universe theories of the origin of the universe do not answer the ultimate origin question. How did the quantum laws that describe the Big Bang–producing vacuum fluctuation originate? Heinz Pagels, one of the pioneers in Vacuum Fluctuation theory, asks this question:

> The nothingness 'before' the creation of the universe is the most complete void that we can imagine—no space, time or matter existed. It is a world without place, without duration or eternity, without number—it is what the mathematicians call 'empty set'. Yet this unthinkable void converts itself into the plenum of existence—a necessary consequence of physical laws. Where are these laws written into that void? What 'tells' the void that it is pregnant with a possible universe? It would seem that even the void is subject to law, a logic that exists prior to space and time.[24]

In his critical study of contemporary cosmology, *God and the Cosmologists*, Stanley Jaki makes a far more telling point: "The method of physics always means an inference from one observable state to another. Observability in turn implies that the thing observed is not absolutely inert and can therefore interact with the observer and its instruments. This is why any talk about an eternal inert universe preceding the actual universe has no place in physics. As for those who speak of esoteric fluctuations in the vacuum prior to the actual physical processes that can be traced back to fifteen billion or so years, they either mean real physical processes or they do not".[25]

The concept of multiple universes does not mitigate the dilemma as Stoeger shows: "If there really are multiple universes, it seems very difficult to see how we would ever find out about the others observationally. But having multiple universes still does not begin to deal with the ultimate questions of why they exist to begin with and why as a collection or ensemble they have the order and potentiality they have for at least one of them to become a universe like ours. It merely postpones these ultimate questions to a previous step. An ensemble of universes by themselves cannot ultimately account for what exists and the character of what exists" (page 267f.).

Perhaps the only other final explanation for the existence of the universe that can be stated in scientific terms would be a theory showing that the universe would necessarily have to be the way it is. Hawking states two reasons why any theory purporting to be comprehensive and necessary in this manner would still be limited:

> Even if we do discover a complete unified theory, it would not mean that we would be able to predict events in general, for two reasons. The first is the limitation that the uncertainty principle of quantum mechanics sets on our powers of prediction. There is nothing we can do to get around that. In practice, however, this first limitation is less restrictive than the second one. It arises from the fact that we could not solve the equations of the theory exactly, except in very simple situations.[26]

At another level, Gödel's Incompleteness Theorem—a theorem that has become all but indisputable since its formulation in 1930—is fatal for any attempt to derive necessarily true theories. Ferris describes the theorem and its implications:

> Gödel's theorem establishes that the full validity of any system, including a scientific one, cannot be demonstrated within that system itself. In other words, the comprehensibility of a theory cannot be established unless there is something outside the frame against which to test—something beyond the boundary defined by a thermodynamics equation, or by the collapse of the quantum wave function, or by any other theory or law. And if there is such a wider reference frame, then the theory by definition does not explain everything. In short, there is not and never will be a complete and comprehensive scientific account of the universe that can be proved valid. The Creator must have been fond of uncertainty, for He (or She) has given it to us for keeps.[27]

No scientific theory, it seems, can bridge the gulf between absolute nothingness and a full-fledged universe (or fledgling universes). This ultimate origin question is a metascientific question—one which science can ask but not answer. This seems to be the point made by Professor Charles Townes: "It is true that physicists hope to look behind the 'big bang' and possibly to explain the origin of our universe as, for example, a type of fluctuation. But then, of what is it a fluctuation and how did this in turn begin to exist? In my view, the question of origin seems always left unanswered if we explore from a scientific view alone. Thus, I believe there is a need for some religious or metaphysical explanation if we are to have one" (page 123).

Recognition that the question of the origin of the universe is a metascientific question does not necessarily entail an end to inquiry because the Principle of Explanation—which we shall refer to here as the PE—is itself a metascientific principle rooted in the essence of rationality. This being the case, the PE can legitimately be extended to metascientific levels of explanation. ("Explanation" is dealt with in Paul Davies's *The Mind of God* [New York: Simon and Schuster, 1992] and in Richard Swinburne's "The Limits of Explanation" [in *Explanation and its Limits*, edited by Dudley Knowles, Cambridge University Press, 1990].)

When applied without reservation, the PE could ground the scientific enterprise in a religious view of reality. Such a connection formed the core of Einstein's views on science and religion: "Certain it is that a conviction, akin to religious feeling, of the rationality or intelligibility of the world lies behind all scientific work of a higher order. . . . This firm belief, a belief bound up with deep feeling, in a superior mind that reveals itself in the world of experience, represents my conception of God".[28] Belief in the "rationality or intelligibility of the world", according to Einstein, is implicitly a belief "in *a superior mind* that reveals itself in the world of experience". As it relates to the ultimate origin of the universe (or universes), the PE path from the "world of experience" to a "superior mind" is usually represented as a cosmological or a teleological argument.

· · · · · · · · · ·

From Cosmos to Theos

Most cosmological and teleological arguments are based on the PE and their validity is bound up with the validity of the PE as an essential and ultimate principle of reality. In its most common forms, cosmological arguments begin with the premiss that any-thing which exists must have an explanation adequate to fully account for its existence either in itself or in something else ("explanation" is understood here in the generic sense of "expla-nation of origin": sophisticated distinctions between regularity explanations and rationality explanations and the attempt to reduce all explanations to personal explanations belong to another level of refinement). From this starting-point, it proceeds to show that "universe" is a term used to refer to all entities that do not have an explanation for their existence in themselves and that the universe is not a super-entity over and above the entities that constitute it. Entities which do not explain their own exist-ence contribute *zero* to their own origin or birth (that is, they are not responsible for their own existence but owe their origin to others). The existence of any one of the entities that make up the universe poses a problem. It has a cause (someone or something else brought it into existence), and this cause is also caused by still another cause, which was also caused by still another cause, and so on.

Logically, there are two possible explanations for the exis-tence of this particular entity or any of the caused entities that constitute the universe. Either there is an infinite series of caused causes or there is an ultimate uncaused cause (a cause which does not owe its origin to any one or anything else and can fully explain its own existence). The existence of an infinite series of caused causes is not an adequate explanation because such a series would still not explain the existence of even one caused entity: every entity in the infinite series contributes zero to its own origin, and an infinite number of zeroes is still zero. Just as an infinite series of mail-carriers cannot explain the existence of one letter, an infinite series of entities which were caused cannot explain the existence they have received. The only viable expla-nation, then, for the existence of any one of the entities or all of the entities that make up the universe would be the existence of an ultimate uncaused being—a being that did not receive exis-tence from anyone or anything else and can completely explain its own existence. This self-explanatory being is commonly called "God" and is the explanatory ultimate demanded by all non-self-explanatory entities from sub-atomic particles to galax-ies. Cosmological arguments do not reason from the fact that everything in the universe has a cause in space and time to the conclusion that the universe has a cause in space and time: these arguments point out, rather, that everything in the universe is non-self-explanatory, which means that the explanation of the universe does not lie in itself but must lie in a self-explanatory being.[29]

Stoeger restates the cosmological argument as it applies to the data of modern science:

> The existence of something, whether it be energy, material particles, or operative laws, requires a cause which either necessarily exists in itself, or ultimately rests on a cause which necessarily exists in itself—a primary cause, the first in the causal chain, not needing any other cause to explain its existence. To have a chain of contingent entities stretching back into the past, none of which explains its own existence and none of which causally depends on an entity or cause which explains its own existence, is simply unintelligible. Or to put it another way, to trace the underlying causes of an entity to more and more fundamental entities and processes which themselves depend on other entities, causes and processes for their existence does not fully explain the existence of the entity in question, unless the search terminates in an entity or process which explains or causes itself, which *necessarily* exists—which of its nature cannot not exist. If it terminates, or seems to terminate, in an entity—a geometric manifold, a symmetry group, some 'cosmic egg'—which is not necessary in this sense and we simply throw up our hands and say that this entity simply exists—simply is—without searching for any primary cause, upon which to rest the 'cosmic egg', then we have prematurely and arbitrarily abandoned our quest for the intelligibility of the whole chain. Then we have no explanation whatsoever for the ultimate cause of the existence of anything—or for why *this* symmetry group or manifold exists, and not some other imaginable one, or why this particular one which we can mathematically investigate is given concretization within reality rather than some other one. If it is not necessary—as symmetry groups and geometric manifolds are not—then there is nothing in its nature which specifies that it must be concretized or instantiated in physical reality. . . . The problem of the origin of the universe, philosophically speaking, is not basically that of a temporal origin, but rather that of a 'causal origin' which renders, first of all, its existence and, secondly, its characteristic order intelligible—which adequately explains or accounts for both" (page 257).

To be sure, cosmological arguments have not convinced everyone. But these arguments seem to derive their validity from the same PE that drives the scientific enterprise.[30] And most of the common objections to these arguments appear to be fallacious.

An infinite series of causes, it has been argued, is not a sufficient explanation for the existence of any one of the causes. In his famous trilogy on the existence of God—*The Coherence of Theism, The Existence of God,* and *Faith and Reason*—Oxford philosopher Richard Swinburne shows that this is the case even if the universe existed over an infinite period of time: "If the universe is of infinite age . . . if the only causes of its past states are prior past states, the set of past states as a whole will have no cause and so no explanation. This will hold if each state has a complete scientific explanation in terms of a prior state, and so God is not involved. For although each state of the universe will have a complete explanation (unlike in the case where the universe is finite, where its first state will not have any explanation), the whole infinite series will have no explanation, for there will be no causes of members of the series, lying outside the series. In that case the existence of the universe over infinite time will be an inexplicable brute fact. There will be an explanation . . . of why, once existent, it continues to exist. But what will be inexplicable is the non-existence of a time before which there was no universe".[31]

Perhaps the most common objection to the cosmological argument is the question, Who created God? Astronomer Carl Sagan, for instance, says, "To my mind, it seems not fully satisfactory to say that there was a first cause. That seems to postpone dealing with the problem rather than solving it. If we say 'God' made the universe, then surely the next question is 'Who made God?' If we say 'God was always here', why not say the universe was always here? If we say that the question 'Where did God come from?' is too tough for us poor mortals to understand, then why not say that the question of, 'Where did the universe come from?' is too tough for us mortals? In what way, exactly, does the God hypothesis advance our knowledge of cosmology?"[32]

Sagan's contention that God is no more an explanation for the origin of the universe than the universe is itself, however, rests on a misconception. If the universe is said to be ultimate, it is inexplicable because its existence *cannot* be explained. If God is taken as ultimate, God's existence *need not* be explained because God is self-explanatory. Only a misunderstanding of the meaning of the concept "God" can lead to questions like "Who made God?" or "Where did God come from?" It is as absurd to ask for an explanation for the existence of a self-explanatory being as it is to ask "Why is a circle round?" The "Who created God?" approach is subjected to further analysis by Professor H.D. Lewis and Professor Hugo Meynell later in this volume (see Part Three). As Meynell says, "Suppose God is . . . that on the understanding and will of which *all* else depends. In that case, God, in virtue of being God, *could not depend on* any other being or beings. The rational theist may thus claim that God is *required* in explanation of the otherwise 'brute facts' of the world, without being such as to require explanation in turn" (page 246).

Meynell's comments are an appropriate rejoinder to another common objection to cosmological arguments: the claim that the universe is just a brute fact for which we should not expect or seek an explanation. In a well-known debate, Bertrand Russell asserted that the universe is there and that's all and we should not ask for an explanation of the existence of the universe. Philosophers like Jean-Paul Sartre have maintained that the universe is "gratuitous", *de trop*, "just there". In this volume, the noted atheist Antony Flew holds that we should take the existence of the universe "and its most fundamental features as the explanatory ultimates" (page 241).

Remarkably, Professor Flew goes on to "confess" that the atheist "has to be embarrassed by the contemporary cosmological consensus. For it seems that the cosmologists are providing a scientific proof of what St. Thomas [Aquinas] contended could not be proved philosophically; namely, that the universe had a beginning. So long as the universe can be comfortably thought of as being not only without end but also without beginning, it remains easy to urge that its brute existence, and whatever are found to be its most fundamental features, should be accepted as the explanatory ultimates. Although I believe that it remains still correct, it certainly is neither easy nor comfortable to maintain this position in the face of the Big Bang story" (page 241). While it is true that the "contemporary cosmological consensus" is an embarrassment for the "universe-as-ultimate-brute-fact" scenario, the most serious roadblock to this scenario is the very structure of the scientific enterprise.

By its very nature, the PE on which science is based cannot rest in brute facts, facts that are unintelligible and non-explanatory. The assertion that the universe is the "ultimate brute fact" is an assertion that there is no explanation for the existence of the universe and that we should not expect any explanation for its existence. But, as Hugo Meynell has shown, "It is an ultimate consequence of our *a priori* assumptions about the nature of the world and of our knowledge of it [*as manifested, for instance, in the scientific method*] . . . that there cannot be 'brute facts'. A putative fact which turns out to be incapable of being fitted into any framework of explanation is not a fact at all; it is impossible to spell out what it would be coherently to suppose, let alone to be assured of, the existence of such a 'fact'. The existence of God would not be a 'brute fact' in the sense objected to, since, as postulated in terms of the form of argument advanced here, God, by his nature, is the sort of being whose understanding and will would explain the how it is and that it is of everything else, without himself being capable of being explained in the same kind of way. That on whose existence, understanding and will the existence of everything else depended could not be dependent on the existence of anything else".[33] There is a qualitative difference between the concepts of God and the universe that is significant in the quest for an explanatory ultimate: God is by definition a Being of infinite intelligence, the source and summit of rationality, whereas the universe in itself is never thought of as intelligent or rational (though it manifests signs of being grounded in intelligence and rationality).[34]

It is often assumed that David Hume and Immanuel Kant have shown that all arguments for the existence of God are necessarily flawed. Hume's and Kant's critiques, however, are derived, to a great extent, from their eccentric theories of knowledge. Jaki, Meynell, and a number of other thinkers have repeatedly demonstrated that the Humean and Kantian theories of knowledge used against theistic arguments cannot be accepted uncritically because they fly in the face of the key assumptions underlying the scientific method. Russell himself once said, "Kant deluged the philosophical world with muddle and mystery, from which it is only now beginning to emerge. Kant has the reputation of being the greatest of modern philosophers, but to my mind he was a mere misfortune".[35] Additionally, Kant's attempt to identify the cosmological argument with the ontological argument has often been effectively derailed, most notably in A.E. Taylor's *Theism*.[36] It cannot be denied that cosmological arguments involve some degree of reasoning from concepts—the same concepts that are fundamental to common sense and science. But these arguments are not entirely or even mainly based on reasoning from concepts: they rest on the solid ground of everyday experience and a universe that calls out for ultimate explanation.

Professor T.D. Sullivan points out that the notion that everything that exists came to be without a cause is an incoherent one if subjected to careful analysis:

> Even if the claim that everything that comes to be could do so without a cause implies no *self*-contradiction, it may contradict something we know. Does it? That depends, of course, on what we know. Suppose we know that at least *one* thing (substance, aggregate, event, property, or action—"existence or mode of existence") has a cause. I will argue that given just this much it is absurd to say everything could come to be without a cause. That is, there is an implicit contradiction in saying, "Of course some things in fact have causes, but everything, including the very things that happen to have causes, could come to be without a cause".[37]

The train of thought that leads from the universe to a transcendent self-explanatory Being is, above all, the response of reason to the facts of experience. "The direction of thought towards the unconditioned (God)", writes Copleston, "is simply the movement of reason itself in its process of understanding in a given context. . . . The completely isolated finite thing is unintelligible, in the sense that reason cannot rest in this idea but strives to overcome the isolation. . . . Some of those who speak of things as being 'gratuitous', *de trop* or 'just there', betray by the very phrases which they use the fact that their reason is not satisfied with the idea of a finite thing as 'just there'". By his refusal to apply the PE to the universe, the atheist "simply puts a bar to the movement of understanding in a certain context because, for reasons which it can be left to others to determine, he does not wish to travel along a path which, as he sees clearly enough, leads in a certain direction".[38] Because it rises in the "movement of reason", rational theism (as opposed to fideism) has been described as the "ultimate rationalism", "the fulfillment of human rationality".

The inseparable connection between science and the PE may raise questions about the relevance of this principle for the world revealed by quantum physics. With such conceptions as the wave-like behavior of particles and the intrinsic inexactitude of measurements in the sub-atomic world, quantum physics may seem to violate some of the fundamental axioms of the scientific enterprise. But three things should be noted about the applicability of the PE in the quantum realm.

It must be understood, first, that quantum theory, like all of science (including the new science of "chaos" that seeks to extend "explanation" even to non-equilibrium complexity), is a search for fundamental laws that describe and explain the workings of the world, an attempt to crack, in Heinz Pagels's phrase, "the cosmic code". The quantum quest is a quest for the ultimate laws that govern the sub-atomic realm as exemplified in quantum field theory by the pursuit of "grand unified field theories". Implicitly, then, quantum theory assumes that reality is intelligible and rational and can be "explained" (even if explanations are stated in terms of statistical laws, such laws hinge on fundamental principles of consistency). Indeed, as Pagels chronicles, "The observed properties of the quantum particles can be precisely described in the language of mathematics, and within that language the notion of symmetry has come to play an increasingly important role. . . . The role of symmetry in describing the properties of quantum particles is central to the entire enterprise of modern physics".[39] "In quantum physics", writes Professor Anton Zee, "symmetry not only tells us about the underlying laws, it also tells us about the actual physical states".[40]

Secondly, the extrapolations and speculative extravaganzas inspired by quantum theory may be misleading. Professor David Bohm, an eminent quantum theorist, points out that "An interpretation, such as the various interpretations of quantum theory, is in no sense a *deduction* from experimental facts or from the mathematics of a theory. Rather it is a proposal of what the theory might *mean* in a physical and intuitively comprehensive sense. . . . In essence, *all* the available interpretations of the quantum theory, and indeed of any other physical theory, depend fundamentally on implicit or explicit philosophical assumptions, as well as on assumptions that arise in countless other ways from beyond the field of physics".[41] Besides, no matter how radically theories like Bohr's Principle of Complementarity and the Many Worlds theories conflict with commonsense conceptions, they are still attempts to describe and *explain* "what is". Professor Bohm suggests that "the quantum theory demonstrates the need for radically new notions of order, and the confusions and failures associated with theory may be due to an attempt to understand something radically new in terms of an older order".[42]

Thirdly, as Professor Jaki explains in his Gifford Lectures, the assumptions of quantum theory "were not deducible from measurements and from the mathematical techniques of matrix and wave mechanics. Such assumptions were the strict, nonstatistical equality of all quanta, of all electrons, of all fundamental particles within a given class, to say nothing of the invariability of the laws of electrodynamics in all atomic processes and measurements". In addition, "the very backbone of matrix mechanics, the universal validity for atomic physics of a noncummutative algebra, was not at all a derivative of observation".[43]

The naive notion that reality disappears when viewed from a quantum perspective was refuted by one of the principal architects of quantum theory, Werner Heisenberg: "The criticism of the Copenhagen Interpretation of the quantum theory rests quite generally on the anxiety that, with this interpretation, the concept of 'objective reality' which forms the basis of classical physics might be driven out of physics. As we have here exhaustively shown, this anxiety is groundless, since the 'actual' plays the same decisive part in quantum theory as it does in classical physics. The Copenhagen Interpretation is indeed based upon the existence of processes which can be simply described in terms of space and time, i.e., in terms of classical concepts, and which thus compose our 'reality' in the proper sense".[44]

Another misleading artifact of quantum speculation is the idea that quantum theory entails the causeless, spontaneous creation of something from nothing. As pointed out earlier, absolute nothingness can never be an object of scientific inquiry. In any case, the nothing in question here turns out either to have antecedents in some minimal something or to actually be "something". To quote Professor Paul Davies: "The processes represented here do not represent the creation of matter out of nothing, but the conversion of pre-existing energy into material form. We still have to account for where the energy came from in the first place. . . . Quantum physics has to exist (in some sense) so that a quantum transition can generate the cosmos in the first place".[45] Professor Adolf Grünbaum warns that some commonly used terms are misleading: "In the case of the so-called 'pair-creation' of a particle and its anti-particle, such as a positron and an electron, their rest-mass formation *as such* occurs by conversion of other forms of energy such as a gamma ray into them. . . . The phrase 'pair annihilation' obscures the fact that the energy of the original positive rest-mass of the particles re-appears in the resulting gamma radiation, although the term 'annihilation-radiation' is not similarly misleading. Corresponding remarks apply to the transformation of gamma radiation into an electron-positron pair: such pair-production is certainly not a case of pair-'creation' *out of nothing*".[46]

It is clear from the above that inquiries in the quantum realm do not eliminate the Principle of Explanation that underlies the scientific enterprise and that leads ultimately to "a superior mind" behind the world of experience. It is significant that pioneers of the quantum quest like Planck and Heisenberg saw no conflict between the scientific enterprise and a religious view of reality (as illustrated by the statements quoted on page 1).

. . . .

Anthropic Arguments

Science, we have seen, cannot deal with absolute nothingness. Science "works" when you have a universe, an intelligible system of processes and laws and entities and structures. Once the ultimate origin question—the origin of being from nothingness—has been resolved in terms of an ultimate metascientific explanation, origin questions that assume a pre-existing "something" causing or evolving into something else (rather than "something" coming from "nothing") can be addressed by utilizing scientific tools and techniques. Thus, the drama of cosmic history that began with the Big Bang and continues to this day can be understood in terms of what Charles Darwin described as "secondary causes"—causes that can be studied purely by means of the scientific method—ultimately deriving from the Creator: "To my mind it accords better with what we know of the laws impressed on matter by the Creator, that the production and extinction of the past and present inhabitants of the world should have been due to secondary causes, like those determining the birth and death of the individual".[47] Traditional theists like Augustine understood the creative process in terms of "causal connections" rather than temporal interventions: "He made that which gave time its beginning, as He made all things together, disposing them in an order based not on intervals of time but on causal connections".[48]

It must be remembered, however, that God's relation to the universe is not just one of creation but is also one of conservation in being. Professor Arthur Peacocke "unpacks" this concept thus: "The cosmos continues to exist at all times by the sustaining creative will of God without which it would simply not be at all. . . . Clearly, if God is 'outside' time in some sense, that is, if time itself is other than God and part of the created cosmos, there is no more difficulty in regarding God as having a creative relationship with the cosmos at all times than postulating a special creative relation only at some posited 'zero' time". [49]

Many contemporary scientists are fascinated by apparent evidence that cosmic processes and patterns are manifestations of an extra-cosmic intelligence. Thus various versions of the Anthropic Principle—the principle that attempts to explain the extraordinary array of cosmic coincidences that made human life possible in terms of a universe "tailor-made" for *Homo sapiens*—function as today's teleological arguments. In *The Cosmic Blueprint*, Professor Paul Davies argues that "The very fact that the universe is creative, and that the laws have permitted complex structures to emerge and develop to the point of consciousness—in other words, that the universe has organized its own self-awareness—is for me powerful evidence that there is 'something going on' behind it all. The impression of design is overwhelming".[50] And, in *Infinite in All Directions*, physicist Freeman Dyson maintains that

> The choice of laws of nature, and the choice of initial conditions for the universe, are questions belonging to meta-science and not to science. Science is restricted to the explanation of phenomena within the universe. Teleology is not forbidden when explanations go beyond science into meta-science. The most familiar example of a meta-scientific explanation is the so-called Anthropic Principle. . . . The last of the five philosophical problems is the problem of final aims. The problem here is to try to formulate some statement of the ultimate purpose of the universe. In other words, the problem is to read God's mind.[51]

Finally, Professor John Leslie, one of today's leading anthropic theorists, addresses the issue of explaining our existence as observers of the universe:

> Let us leave aside for the moment any struggles with whether the bare existence of any world at all cries out to be explained, or again, all fuss over whether an explanation is needed for the sheer fact that the world obeys causal laws. Let us concentrate on whether observership needs explanation. Modern cosmology appears to confirm that it does. And though the Many Worlds hypothesis . . . provides a possible explanation here, it is hardly one noteworthy for its simplicity, is it? . . . the God hypothesis could well be the more reasonable. And in that case to say there was no actual evidence for it would be like saying

that because black holes cannot be observed directly there could be no signs of their existence".[52]

From an anthropic standpoint, the key question in origin of life studies is not that of specific supernatural intervention in the processes that culminated in life but of the Design and Intelligence implied by these processes. Even if or whether there is a consensus in the scientific community on the mechanisms responsible for the origin of life, the link between the origin of life and the ultimate origin question cannot be ignored. Self-organizing scenarios for the origin of life do not obviate the need for an ultimate explanation. If matter or energy has self-organizing tendencies, we are still left with two questions: How did matter and energy come to have these tendencies? and how did the matter and energy that have these tendencies come into being? Similar questions can be asked about the popular view that the universe is such that it will inevitably give rise to life: How did the universe come to have this capability? and how did a universe with such a capability come to be?

The origin of *Homo sapiens* poses problems similar to the questions of the origin of life and also some that are qualitatively different. On a biological level, most researchers have fairly definite ideas about the "genealogy" of *Homo sapiens*; unfortunately, almost all the work done on the hominids and on evolutionary theory in general remains at a biological level. The problem of consciousness—the problem of the human mind—is rarely if ever addressed. In this anthology, however, this metascientific problem is taken seriously by a few of the contributors who consider *Homo sapiens* in the light of the mental activity that is a fundamental feature of human experience.

Professor George Snell points out that "Matter and consciousness appear to me to be distinct though tightly linked entities. Consciousness we know directly only through our personal, inner sensations, but as I see it, and as I assume other people see it, it is quite unlike the material world. . . . While science can study the nature of the material world, and is the appropriate tool for the study of extrasensory perception, it can tell us essentially nothing as to the nature of consciousness" (page 210f.). Sir Nevill Mott argues that "There is one 'gap' for which there will never be a scientific explanation, and that is man's consciousness. No scientist in the future, equipped with a super-computer of the twenty-first century or beyond, will be able to set it to work and show that he is thinking about it. This has been argued by my successor as the head of the Cavendish Laboratory at Cambridge, Sir Brian Pippard, in an essay entitled 'The Invincible Ignorance of Science' (*Contemporary Physics* [Taylor and Francis, London] volume 29, p. 393, 1988). Pippard, an agnostic about God, does not describe this as the 'gap' where God makes himself known. But I would deduce from this hypothesis, that the way God plays a part in our lives is because countless men and women claim to be conscious of Him, when they seek Him, and accept that He is the God of love. God can speak to us and show us how we have to live".[53]

According to Sir John Eccles the origin of consciousness is relevant to the origin of *Homo sapiens*: "The only certainty we have is that we exist as unique self-conscious beings, each unique, never to be repeated. This I regard as outside the evolutionary process. The evolutionary process gives rise to my body and brain but, dualistically speaking, that is one side of the transaction. The other side is my conscious being itself in association with the brain for this period when I have a brain on this earth. So that brain and body are in the evolutionary process but yet not fully explained in this way. But the conscious self is not in the Darwinian evolutionary process at all. I think it is a divine creation. . . . And this is a loving creation. You have to think of it as not just by a Creator Who tosses off souls one after the other. This is a loving Creator giving us all these wonderful gifts" (page 163f.). As described in a primal metaphor as fundamental as the concept of *creatio ex nihilo*, the human mind is the *imago dei*.

· · · · · · · · · ·

Notes

1. Albert Einstein, *Lettres à Maurice Solovine* reproduits en facsimile et traduits en français (Paris: Gauthier-Villars, 1956), pp. 102–3.
2. Albert Einstein, *Ideas and Opinions*, trans. Sonja Bargmann (New York: Dell Publishing Company, 1973), p. 255.
3. Max Planck, *Where is Science Going?*, trans. with biographical note by James Murphy (New York: W.W. Norton, 1977), p. 168.
4. Werner Heisenberg, *Across the Frontiers*, trans. Peter Heath (San Francisco: Harper and Row, 1974), p. 213.
5. Stephen Hawking, *A Brief History of Time* (New York: Bantam, 1988), p. 175.
6. Quoted in Timothy Ferris, *Coming of Age in the Milky Way* (New York: William Morrow, 1988), p. 177.
7. Stephen Hawking, *A Brief History of Time*, p. 175.

8. Robert Jastrow, *God and the Astronomers* (New York: W.W. Norton, 1978), p. 15.

9. Jaegwon Kim, "Explanation in Science," *Encyclopedia of Philosophy*, Volume Three (New York: Macmillan, 1967), p. 159.

10. Hugo Meynell, *The Intelligible Universe* (Totowa, New Jersey: Barnes and Noble, 1982), p. 84.

11. Albert Einstein, *Ideas and Opinions*, p. 49.

12. Frederick Copleston, *Religion and Philosophy* (Dublin: Gill and Macmillan, 1974), p. 174.

13. *Leibniz Selections*, edited by Philip P. Wiener (New York: Charles Scribner's Sons, 1951), p. 527.

14. Stephen Hawking, *A Brief History of Time*, pp. 33-34.

15. Stanley Jaki, *The Absolute Beneath the Relative* (Lanham, Maryland: University Press of America, 1988), p. 11.

16. Heinz Pagels, *Perfect Symmetry* (New York: Bantam, 1985), p. 146.

17. "Universal Truths", *Scientific American*, October, 1990.

18. Stephen Hawking, *A Brief History of Time*, p. 173.

19. Stephen Hawking in *Origins: The Lives and Worlds of Modern Cosmologists*, edited by Alan Lightman and Roberta Brawer (Cambridge: Harvard University Press, 1990), p. 397.

20. Stephen Hawking, *A Brief History of Time*, p. 174.

21. Roger Penrose in *Origins: The Lives and Worlds of Modern Cosmologists*, edited by Alan Lightman and Roberta Brawer, (Cambridge: Harvard University Press, 1990), p. 433.

22. Don N. Page, "Hawking's Timely Story," *Nature*, April, 1988, p. 742-43. The philosophical problems generated by Hawking's cosmological speculations are further scrutinized by Professor William Craig in *The British Journal for the Philosophy of Science*.

23. Timothy Ferris, *Coming of Age in the Milky Way*, pp. 358-59.

24. Heinz Pagels, *Perfect Symmetry*, p. 365.

25. Stanley Jaki, *God and the Cosmologists* (Washington, D.C.: Regnery Gateway, 1989), p. 81. Jaki's critical distinctions indicate that the problem of the origin of the universe is a metascientific and not a scientific problem. In this respect, Professor Adolf Grünbaum is right in talking about "The Pseudo-Problem of Creation in Physical Cosmology" (*Philosophy of Science*, September, 1989). But what is not a "problem" in physical cosmology does not cease to be a problem in ontological inquiry.

26. Stephen Hawking, *A Brief History of Time*, p. 168.

27. Timothy Ferris, *Coming of Age in the Milky Way*, p. 384.

28. Albert Einstein, *Ideas and Opinions*, p. 255.

29. Strictly speaking, as Professor Meynell has pointed out, there is one other possibility: a multitude of uncaused causes. E.L. Mascall and H.D. Lewis, among others, have shown why this possibility does not resolve the issue of ultimate explanation. Mascall: "It might, how-

ever, be replied (1) that the supposition that there are two or more self-existent beings each of which is the transcendent cause of a different set of finite beings (or perhaps different sets of which are the transcendent causes of different sets of finite beings) leaves their own co-existence unexplained, (2) that the constitution of finite beings with their transcendent cause manifests the latter as being absolutely ultimate and not as one who shares his ultimacy with others. Such replies as these receive welcome support from considerations of the nature of morality." *The Openness of Being* (Philadelphia: Westminster Press, 1971), p. 119. Lewis: "[The Ultimate Reality] can clearly not be one among many of its kind. It must be beyond the relatedness of other things and retain in itself the principle by which other entities stand in relation to one another. If it had a duplicate the relation of the two realities to one another would have to be understood again in a way that pointed beyond both to some more ultimate and radically different sort of being which is itself limited by nothing. God is not *primus inter pares* but absolute." *Philosophy of Religion* (London: English Universities Press, 1965), p. 147.

30. Some critics have impugned the validity of a Principle of Explanation on the grounds that it leads to incoherence when confronted with the question of ultimate explanation. The problem with a Being that necessarily exists as an explanation for the universe, it is alleged, is the relation between the universe and the Necessary Being: if the universe-explaining Being is necessary, then so is the universe and its relation to the Necessary Being. In response to these charges, it must first be pointed out that even if alleged problems with the concept of a Necessary Being can be shown to be legitimate, the invalidation of this concept is not simultaneously a disproof of a Principle of Explanation. We know that the Principle "works" in everyday contexts, and its validity in these contexts is not in question because of apparent problems in applying it in ultimate contexts. Secondly, as Eleonore Stump and Norman Kretzmann have argued, the concept of Divine Simplicity entails "that God is a logically necessary being all of whose acts of will are at least conditionally necessitated, and among these acts of will is the volition that certain things be contingent." "Absolute Simplicity", *Faith and Philosophy*, Volume 2, Number 4, p. 377.

31. Richard Swinburne, *The Existence of God* (Oxford: Clarendon Press, 1979), p. 124.

32. "God and Carl Sagan: Is the Cosmos Big Enough for Them?," *U.S. Catholic*, May 1981, p. 20.

33. Hugo Meynell, *The Intelligible Universe*, pp. 104–5.

34. Stump and Kretzmann show clearly why the existence of God is not a "brute fact" like the existence of the universe: "The answer to the question 'Why does God exist?' is that he cannot not exist, and the

reason he cannot not exist is that because he is absolutely simple he is identical with his nature. If his nature is internally consistent, it exists in all possible worlds, and so God, identical with his nature, exists in all possible worlds. The necessity of God's existence is not one more characteristic of God which needs an explanation of its own but is instead a logical consequence of God's absolute simplicity". "Absolute Simplicity", p. 377.

35. Bertrand Russell, *An Outline of Philosophy*, (London: Allen and Unwin, 1927), p. 83.

36. *Encyclopedia of Religion and Ethics*, Volume 12, p. 278.

37. "Coming to be Without a Cause," *Philosophy*, Volume 65, 1990, p. 262.

38. Frederick Copleston, *Religion and Philosophy*, pp. 176–77.

39. Heinz Pagels, *Perfect Symmetry*, pp. 176–77.

40. Anton Zee, *Fearful Symmetry* (New York: Macmillan, 1986), p. 150.

41. David Bohm, *Science, Order and Creativity* (New York: Bantam, 1987), p. 102.

42. David Bohm, *Science, Order and Creativity*, p. 103.

43. Stanley Jaki, *The Road of Science and the Ways to God* (Chicago: University of Chicago Press, 1978), p. 211.

44. Werner Heisenberg, "The Development of the Interpretation of the Quantum Theory," *Niels Bohr and the Development of Physics*, edited by W. Pauli, L. Rosenfeld, and V. Weisskopf (London: Pergamon Press, 1955), p. 28.

45. Paul Davies, *God and the New Physics* (New York: Simon and Schuster, 1983), pp. 31, 217.

46. Adolf Grünbaum, "The Pseudo-Problem of Creation," *Philosophy of Science*, September, 1989, pp. 383–85.

47. Charles Darwin, *The Origin of Species by means of Natural Selection*, (London: Penguin, 1968), p. 460.

48. St. Augustine, *The Literal Meaning of Genesis*, Volume I, trans. and annotated by John Hammond Taylor, S.J. (New York: Newman Press, 1982), p. 154.

49. A.R. Peacocke, *Science and the Christian Experiment* (London: Oxford University Press, 1971), pp. 120, 123.

50. Paul Davies, *The Cosmic Blueprint* (New York: Simon and Schuster, 1988), p. 203.

51. Freeman Dyson, *Infinite in All Directions* (New York: Harper and Row, 1988), pp. 296-98.

52. John Leslie, "Observership in Cosmology: the Anthropic Principle," *Mind*, 1983, Volume XCII, pp. 573–79.

53. Sir Nevill Mott, "Science and the Existence of God," *Cosmos, Bios, Theos*, page 66. In *The Emperor's New Mind*, Professor Roger Penrose picks apart the notion that computer systems can duplicate the mind and shows that human creativity and consciousness are, in principle, beyond the reach of even the most sophisticated computers. John Lucas used Gödel's theorem to reach this conclusion in *The Freedom of the Will*.

PART ONE

· · · · · · · · · ·

Astronomers, Mathematicians,
and Physicists

1

Who Arranged for These Laws to Cooperate So Well?

Professor Ulrich J. Becker

• • • • • • • • • •

- Born 17 December 1938

- Ph.D. in physics, University of Hamburg, 1968

- Currently Professor of Physics, Massachusetts Institute of Technology, and Member of the Research Council of Europe in Geneva, Switzerland

- Areas of specialization and accomplishments: high-energy particle physics; co-discoverer of the new meson J (3.1); study of the variety of new particles, particularly vector mesons

- Professor Becker on:

    the origin of the universe: "Scientifically: unknown".

    the origin of life: "Even if *new* 'chaotic-to-order' models enhance the probability by many orders of magnitude to form the first reprodicing entity, the question of the origin is not answered without addressing who arranged for these laws to cooperate so well".

    the origin of *Homo sapiens:* ". . . Taungchild, Neanderthal, Cro-Magnon. . ."

    God: "How can I exist without a creator? I am not aware of any compelling answer ever given".

• • • • • • • • • •

1 What do you think should be the relationship between religion and science?

Not one of conflict. There is no room for human arrogance or intolerance in this relation. At all times scientists should be aware of the incompleteness of their knowledge and attempts, as just the greatest among them were.

2 What is your view on the origin of the universe: both on a scientific level and—if you see the need—on a metaphysical level?

> Scientifically: unknown. Speculations are compatible with: "In the beginning there was light!" namely (heavy photons, Z's, and so on—force carrier bosons). Ironically we consist of the condensed matter made of zerospace (pointlike) fermions. Is it only the Pauli exclusion that provides our space, and the Higgs fields our energy? Anyhow don't be afraid to call it creation.

3 What is your view on the origin of life: both on a scientific level and—if you see the need—on a metaphysical level?

> It happened 3×10^9 years ago. Luria and Ziegler argue on plausibilistic levels to overcome the minute probability of an accidental start as scientifically rigorous thermodynamics predicts. Even if *new* "chaos-to-order" models enhance the probability by many orders of magnitude to form the first reproducing entity, the question of the origin is not answered without addressing who arranged for these laws to cooperate so well.

4 What is your view on the origin of *Homo sapiens*?

> Who was the first one? From all that we presently know: Taung-child, Neanderthal, Cro-Magnon, Akkadians? Or better—the first one to excel mere opportunism by recognizing himself with "good and evil". *Sapiens* means "wise"—are we qualified?

5 How should science—and the scientist—approach origin questions, specifically the origin of the universe and the origin of life?

> Honestly, with clear separation of scientific knowledge from plausible or pretentious speculation. If you discovered how one wheel in the "clock" turns—you may *speculate* how the rest move, but you are not entitled to call this scientific and better leave alone the question of who wound up the spring.

> (He who cannot—for ego-cosmetics or other reasons—distinguish fact from speculation, please stay off TV and public debates.)

6 Many prominent scientists—including Darwin, Einstein, and Planck—have considered the concept of God very seriously. What are your thoughts on the concept of God and on the existence of God?

> How can I exist without a creator? I am not aware of any compelling answer ever given. The eighteenth and nineteenth century

determinism (including my dearest Immanuel Kant) tried a universe running down like a clock with a "departed" watchmaker. How naive can one get to use a time argument (departed) for the creator of space-time!

A final remark: If you accept God as being "the Law" (physical, chemical, social, mathematical, moral-ethical, psychological . . .) all points 2, 3, 4, 5, and 6 fall in place.

2

The Origin of the Universe Is, and Always Will Be, a Mystery

• • • • • • • • • •

Professor Stuart Bowyer

- Born 2 August 1934

- Ph.D. in physics, Catholic University, 1965

- Currently Professor of Astronomy, University of California, Berkeley

- Areas of specialization and accomplishments: galactic and extra-galactic X-rays; extreme ultraviolet radiation in earth's atmosphere and from astronomical objects; at Berkeley, he has developed a group which is primarily involved with X-ray, extreme ultraviolet, and far ultraviolet astronomy, and related topics in high energy astrophysics; awarded several patents for space instrumentation and received the Humboldt Foundation of Germany Senior Scientist Award (1982)

- Professor Bowyer on:

  the origin of the universe: "Ultimately, the origin of the universe is, and always will be, a mystery. Science has pressed the level of what can be explained further and further into the early universe, but the mystery is nonetheless there".

  the origin of life: ". . . based on a series of chemical reactions . . ."

  the origin of *Homo sapiens:* ". . . one product of the effects of random genetic variation and natural selection".

  God: "I, at most, am an agnostic".

• • • • • • • • • •

1 What do you think should be the relationship between religion and science?

Physical science should, and does, reflect or describe the physical world. Religion should, and does, reflect or guide the world of human values, both intrinsically and in ways which may possibly be related to value entities outside of usual human experience. Hence religion is forced to take the science description of the world as it is. And physical science has no claim for recognition on the topic of human values.

2 What is your view on the origin of the universe: both on a scientific level and—if you see the need—on a metaphysical level?

> Ultimately, the origin of the universe is, and always will be, a mystery. Science has pressed the level of what can be explained further and further into the early universe, but the mystery is nonetheless there.

3 What is your view on the origin of life: both on a scientific level and—if you see the need—on a metaphysical level?

> The origin of life is based on a series of chemical reactions which, given the proper conditions, will surely occur given enough time.

4 What is your view on the origin of *Homo sapiens*?

> *Homo sapiens* is one product of the effects of random genetic variation and natural selection.

5 How should science—and the scientist—approach origin questions, specifically the origin of the universe and the origin of life?

> Science should approach the questions of the origins of life and the universe as they do any other scientific question.

6 Many prominent scientists—including Darwin, Einstein, and Planck—have considered the concept of God very seriously. What are your thoughts on the concept of God and on the existence of God?

> I believe the concept of God varies widely. Most formulations are not viable in my opinion; a few may be. But I, at most, am an agnostic.

3

Appeal to God May Be Required to Answer the Origin Question

. . . .

Professor Geoffrey F. Chew

- Born 5 June 1924

- Ph.D. in physics, University of Chicago, 1948; Hughes Prize, American Physical Society, 1962

- Currently Dean of Physical Sciences, University of California, Berkeley

- Areas of specialization and accomplishments: developed bootstrap theory of nuclear particles based on analytic S-Matrix; studies on Regge pole theory of high-energy nuclear reactions; works include *S-Matrix Theory of Strong Interactions,* 1961

- Professor Chew on:

    the origin of the universe: "Berkeley quantum cosmologists are seeking a mathematical model that will cover transition from pre-Big Bang (no space-time) to the present epoch".

    the origin of life and of *Homo sapiens*: " 'Life,' including *Homo sapiens,* is only one aspect of structure evolving as the universe expands. . . . Nevertheless, an interesting question is whether 'conscious life' capable of 'free will' involves aspects of the universal quantum state that fail to show up in the semiclassical concept of 'particle' ".

    God: "Appeal to God may be needed to answer the 'origin' question: 'Why should a quantum universe evolving toward a semiclassical limit be consistent?' "

. . . .

My fifty years of exposure to physics and physicists have led to a line of research which requires attention to the nonphysical questions you have posed. Let me begin by telling you something about a theory which several of us here in Berkeley are trying to construct. It falls into a category called (e.g. by Gell-Mann, Hartle, and Hawking) "quantum cosmology". Although both of the separate words "quantum" and "cosmology" are used by people who practice "physical science", the combination leads to questions that are not "scientific", at least in the Galilean sense. Quantum cosmology constitutes a new category of human exploration.

The Copenhagen Interpretation of quantum mechanics is inadmissible when the whole universe is considered; one must find a quantum-mechanical origin for the predominantly "classical" nature of our present universe—which in many respects is describable in terms of localized "matter" carrying energy and momentum. This "physical" viewpoint, which allows one to speak of "observables", "observers", and "observations", is viable only because our present universe is very large and very dilute. From a quantum cosmological viewpoint such notions lack a priori status; their meaning has become possible through evolution of the universe from an epoch of high density (and small dimension) to which no classical ideas pertain. The accuracy (hence, utility) of classical concepts derives from huge dimensionless numbers that characterize the present universe, such as the ratio of universe age (10^{10} years) to Planck time (10^{-44} sec).

Berkeley quantum cosomologists believe even the meaning of space-time to be no more than an asymptotic approximation in the foregoing sense. We agree with Mach that local space-time achieves its meaning from matter (that is, from localized particles that carry energy), and we interpret "Big Bang" as a quantum-dynamical condensation leading to (semiclassical, localized) particles. In our pre–Big Bang epoch there was no meaning for space-time—no meaning for any subdivision of a single universe. Universe expansion subsequent to Big Bang has led to successive local condensations at expanding scales of dimension that allow "objectivity" to achieve meaning. It is nevertheless only an approximation to attribute "identity" to any piece of the universe—to any "object" (including a human being). From a quantum standpoint the entire universe is interconnected; there is a single universal quantum state.

Berkeley quantum cosmologists are seeking a mathematical model that will cover transition from pre–Big Bang (no space-time) to the present epoch. We are striving to represent an asymptotic tendency of quantum-dynamical evolution to approach a semiclassical limit. We hope isolation of local particle collections from the rest of the universe will be recognizable as an approximation that asymptotically increases in accuracy—thereby providing a basis not only for Galilean science but for the language with which your nonscientific questions were posed.

Although I am unsure that quantum cosmology is a sound track, I find it more satisfying than other cosmologies to which I have been exposed. So I here shall react to your questions from the quantum cosmological viewpoint, assuming asymptotic tendency toward a zero-density semiclassical universe (flat spacetime). To achieve such asymptotics a very special quantum state is required, together with a very special evolution generator. The puzzle of quantum cosmology is: how many *different* quantum regimes can lead to semiclassical asymtopia? To find even one is difficult and not yet achieved. I personally incline to believe there is only one choice—in other words, that we are situated within a "bootstrap" universe—determined uniquely by the requirement of consistency.

Now to your questions: The foregoing quantum-evolutionary view removes need to distinguish sharply between "living" and "nonliving" matter. "Life", including *Homo sapiens*, is only one aspect of structure evolving as the universe expands. (Dyson, by the way, has speculated about life in a distant future where DNA, or in fact any molecule, is no longer sustainable.) Nevertheless, an interesting question is whether "conscious life" capable of "free will" involves aspects of the universal quantum state that fail to show up in the semiclassical concept of "particle". The content of our quantum state is not entirely representable through particles. Suppose the missing quantum component is involved in the phenomena of "consciousness" and "free will". To the extent that the laws of physics are completely expressible through the "particle" notion, one would thereby understand why free will lies outside physics.

A related feature of our quantum cosmology is that its strict determinism is not manifested through "time"; the ordinary meaning of time emerges as an asymptotic semiclassical approximation connected to energy (carried by particles). Our quantum determinism is expressed through a dimensionless evolution parameter, whereas "time" is defined as the monotonically-increasing expectation value of a certain operator with dimension $+1$ in an asymptotic domain where fluctuations of the operator become small. The fluctuations, however, are not zero so we do not have strict "time" determinism. To the extent that "observers" are "conscious" of "time" (and not of the dimensionless quantum-evolution parameter), "free will" acting in conjunction with a "physical universe" may be compatible with a deterministic quantum cosmology.

The vagueness of the foregoing reflects our theory's preliminary stage of development. We are presently lacking a theoretical representation of physical measurement—explaining how, in a large and dilute universe, collections of particles and particle interactions may become approximately isolated—in the pattern described by the Galilean-Copenhagen language of "observable", "observer", and "observation". (Implicit in these words is observer "consciousness".)

Appeal to God may be needed to answer the "origin" question, "Why should a quantum universe evolving toward a semiclassical limit be consistent?" (I doubt that consistency will be established through mathematics.)

4

How Is It Possible to Exclude Action Coming from a Transcendent Order of Being?

• • • • • • • • • •

Professor Alexandre Favre

- Born 23 February 1911

- Dr. es Scis., University of Paris, 1937; Prix Marquet, Academy of Sciences; additional awards include: Officier Légion d'honneur; Officier Ordre national du Mérite; Commander des Palmes académiques

- Currently Director of the Institut de Mécanique Statistique de la Turbulencé, Université d'Aix-Marseill

- Areas of specialization and accomplishments: inventor, hypersustentation by moving wall (1934); hyperconvention by moving wall (1951); centrifugal sub-trans-supersonic compressor (1940—used for aircraft jet engines since 1944); apparatus for statistical measurement of correlation coefficient with time delay (1942); appliance for detection of random noise from periodic signals by autocorrelation (1952); initiator space-time correlation measurements in turbulent flows (1942); research on development of statistical equations for turbulent compressible gas (1948)

• • • • • • • • • •

The sciences are rational disciplines for objective knowledge based on the observation of repeatable and verifiable phenomena in order to group them under common laws. But their domains are limited; when problems are solved, others appear. The process diverges, showing that the sciences are incapable of reaching total and ultimate reality.

Philosophy and religion may overstep the bounds of the sciences but must take account of scientific knowledge. The sciences, then, have a negative and a positive role. The negative role is catharsis, to eliminate in philosophy assumptions which contradict scientific observations and to eliminate in religion superstitions and idolatries. The positive role is to disclose the organization of nature, to try to explain it rationally, and to give new general ideas.

In a scientific and philosophical reflection *De la Causalité à la Finalité à propos de la Turbulence* (A. Favre, H. and J. Guitton, A. Lichnerowicz, E. Wolff, eds., Paris: Maloine, 1988), we examined the behavior of the atmosphere-hydrosphere-life ecosystem. This ecosystem exhibits an organization with regulation, optimization, and convergence in favor of life, which is characteristic not only of *causality* but also of teleonomy.

There are inadequacies with explanations of these observations by chance, by creation of order through disorder, by the sole adaptation of living organisms to their medium, and by the extrapolation of determinism to describe both the universe and human actions.

There is compatibility between determinism and necessity, contingency, liberty, complexity. Turbulent flows are deterministic and have regulating mean effects. Teleonomy is the global cause of the convergence of the causes.

It seems to me that three main types of explanations should be considered.

Some philosophical theories affirm the necessity of causalist laws of nature, excluding teleonomy and all action other than that of these laws or any action coming from a transcendent order of being , and affirm the self-organization of matter. I think that matter, inanimate matter or living substance, exhibits behavior that is consistent with the laws of nature indeed. But we have found that these laws have causalistic *and teleonomic* characteristics. It is difficult for me to believe that living organisms, especially the simplest, are able to exhibit teleonomic behavior purely by following causalist laws by themselves. It is even more difficult for me to believe that non-living matter—such as air and water—is able to exhibit teleonomic behavior in favor of life in the atmosphere-hydrosphere system by itself. As for actions coming from a transcendent order, how is it possible to exclude them without ignoring the limitations of the world?

Other theories affirm the necessity (consistent with contingency and liberty) of the laws of nature, causalist and teleonomic, in favor of life, without any action coming from a transcendent order of being, from the "outside world", and affirm the existence of spirituality. I think that spirituality does exist in the world. It appears first in the human spirit which is the manifestation of self-consciousness. But how can we exclude a priori any other spirit than the human spirit while we are unable to understand wholly the organization of the universe? These theories lead to pantheism, materialist or spiritualist. But again, how is it possible to exclude any action coming from a transcendent order of being, from the "outside world", without ignoring the limitation of the world?

Other theories affirm the necessity (consistent with contingency and liberty) of the laws of nature, causalist and teleonomic, in favor of life and affirm the existence of a transcendent organizing power. These theories are not inconsistent with scientific observations and seem to me more satisfactory. The idea of the existence of a transcendent power is a great mystery, but are not the sciences permanently at the borders of mysteries that they solve progressively, without ever being able to reach ultimate reality?

To answer the question asked about the relationship between religion and science, I have not found in my scientific experience anything that contradicts religious faith. Indeed the observation of teleonomy internal to the various sciences has confirmed such faith.

5

Where Matter and Consciousness Came from Is Unknown

• • • • • • • • • •

Professor John Erik Fornaess

• Born 14 October 1946

• Ph.D. in mathematics, University of Washington, 1974

• Currently Professor of Mathematics, Princeton University

• Professor Fornaess on:

the origin of the universe: "The physics of the first fraction of the first second is unknown. Where matter came from is also unknown. It is also unknown where consciousness came from".

the origin of life and of *Homo sapiens*: "The origin of life came about under favorable chemical conditions. . . . Human beings arose this way as well".

God: "I believe that there is a God and that God brings structure to the universe on all levels from elementary particles to living beings to superclusters of galaxies".

• • • • • • • • • •

1 What do you think should be the relationship between religion and science?

I believe that religion should play no role in the development of science. Conversely science should play no essential role in religion.

2 What is your view on the origin of the universe: both on a scientific level and—if you see the need—on a metaphysical level?

The origin of the universe was via the Big Bang some fifteen billion years ago. The physics of the first fraction of the first second is unknown. Where matter came from is also unknown. It is also unknown where consciousness came from.

3 What is your view on the origin of life: both on a scientific level and—if you see the need—on a metaphysical level?

4 What is your view on the origin of *Homo sapiens*?

> The origin of life came about under favorable chemical conditions. Lumps of matter which developed and divided into equal lumps were formed. Chance mutations developed more advanced forms. Human beings arose this way as well.

5 How should science—and the scientist—approach origin questions, specifically the origin of the universe and the origin of life?

> We don't have any idea where the basic ingredients of the universe came from. The best approach to this is probably physics.
>
> The origin of life is less mysterious. I don't think consciousness can be explained as arising from matter.

6 Many prominent scientists—including Darwin, Einstein, and Planck—have considered the concept of God very seriously. What are your thoughts on the concept of God and on the existence of God?

> I believe that there is a God and that God brings structure to the universe on all levels from elementary particles to living beings to superclusters of galaxies.

6

God Is a Characteristic of the Real Universe

. . . .

Professor Conyers Herring

- Born 15 November 1914

- Ph.D. in physics, Princeton University, 1937; awards include Oliver E. Buckley Solid State Physics Prize of the American Physical Society (1959); Von Hippel Award of the Materials Research Society (1980); and the Wolf Prize in Physics (1985)

- Currently Professor Emeritus of Applied Physics, Stanford University

- Areas of specialization include theoretical research on electronic and atomic properties of solids

- Professor Herring on:

  the origin of the universe: "Speculations and models having to do with the origin of the universe have a very legitimate place in science".

  the origin of life and of *Homo sapiens*: ". . . we should not expect our present thinking to parallel that of the writers of ancient religious scriptures, who were interested in getting at truths of a very different kind, more closely related to everyday life".

  God: "Things such as truth, goodness, even happiness, are achievable by virtue of a force that is always present, in the here and now and available to me personally".

. . . .

· **1** What do you think should be the relationship between religion and science?

Religion provides, or should provide, the framework and perspective for all aspects of living. Thus it should guide a scientist's attitudes toward scientific truth, "scientific method", scientific ethics, and so forth. For the layman its role in these matters is more limited, but even the layman needs to have some sort of orientation toward the significance of knowledge and the role of science. But the role of religion, unlike that of science, extends beyond the merely intellectual into the affective and emotional aspects of living.

2 What is your view on the origin of the universe: both on a scientific level and—if you see the need—on a metaphysical level?

Speculations and models having to do with the origin of the universe have a very legitimate place in science, since they can help us to conceptualize the universe we observe in an efficient manner and make useful predictions about it. However, questions like, Did God exist prior to the Big Bang? are meaningless.

3 What is your view on the origin of life: both on a scientific level and—if you see the need—on a metaphysical level?

Not being a biologist or even a chemist, I won't try to express an opinion on the question whether or not any as yet unknown physical laws or mechanism may have to be invoked to account for the origin of life. However, I think it is perfectly meaningful and worthwhile to argue about the mechanisms involved, since any insights we get might guide us in establishing future relations with other "living" beings elsewhere in the universe. Certainly we should not expect our present thinking to parallel that of the writers of ancient religious scriptures, who were interested in getting at truths of a very different kind, more closely related to everyday life.

4 What is your view on the origin of *Homo sapiens*?

My attitudes on this question are similar to those on Question 3 above.

5 How should science—and the scientist—approach origin questions, specifically the origin of the universe and the origin of life?

Again, I think I've largely covered this under Question 3. In general I think the scientist should seek a picture that is as intellectually harmonious as possible and that is as effective as possible in predicting and controlling natural phenomena in the present and future.

6 Many prominent scientists—including Darwin, Einstein, and Planck—have considered the concept of God very seriously. What are your thoughts on the concept of God and on the existence of God?

> We live in a hard , real universe to which we have to adapt. God is a characteristic of that universe—indeed, a miraculous characteristic—that makes that adaptation possible. Things such as truth, goodness, even happiness, are achievable, by virtue of a force that is always present, in the here and now and available to me personally. I reject the idea of a God who long ago set a great clockwork in motion and has since been sitting back as a spectator while mankind wrestles with the puzzle. One reason for my rejection of this is that my scientific experience gives me no reason to believe that there is any "clockwork" model of the universe that is ultimately and finally the correct one. Our scientific theories always work with such models, and they will always be capable of greater and greater refinement, but I feel sure they will always prove imperfect. It is safer, I think, to have faith in the living force that makes this improvement always possible.

7

What Forces Filled the Universe with Energy Fifteen Billion Years Ago?

• • • • • • • • • •

Professor Robert Jastrow

- Born 7 September 1925

- Ph.D. in theoretical physics, Columbia University, 1948

- Currently Professor of Earth Sciences, Dartmouth College, and Director, Goddard Institute of Space Studies

- Areas of specialization and accomplishments: study of nuclear and atmospheric physics; physics of moon and terrestrial planets; works include *Exploration of Space*, 1960; *Red Giants and White Dwarfs*, 1967; *God and the Astronomers*, 1978

- Professor Jastrow on:

    the origin of the universe: "The fact that the universe was once in a dense, hot state is considered proven by almost every scientist".

    the origin of life: "Everything I have studied . . . adds up to a plausible explanation as to how life could have sprung from non-living matter. . . . But that's my opinion. Nobody has demonstrated that life, even a simple bacterium, can evolve from a broth of molecules. . . . as a result of my scientific studies, I have a deeply rooted belief in the philosophies of reductionism and materialism—the view that the whole is equal to the sum of the parts—even with respect to questions relating to life and mind".

    the origin of *Homo sapiens:* "Advanced thinkers in theology—I think the majority—take the view that no contradiction exists whatsoever between their faith and the facts of evolution".

    God: "I suppose a series of miracles could demonstrate to me that forces are at work in the universe that are utterly outside the reach of human comprehension and that is as close as I can come to the proposition of a proof for the existence of a Creator".

• • • • • • • • • •

1 On what grounds do you consider the Big Bang theory a more plausible explanation for the origin of the universe than the Oscillating Universe and the Steady State theories?

The fact that the universe was once in a dense, hot state is considered proven by almost every scientist, through the discovery of the primordial fireball radiation and the measurement of the spectrum—the distribution of intensity with frequency—which is exactly what you would expect of the radiation emanating from the fireball of an explosion.

That fireball radiation proves that the universe was once in an explosive state, and this proves that the Steady State theory is not correct.

The Oscillating Universe theory is consistent with the fireball radiation, because it is possible that the universe collapses and explodes repeatedly. But the evidence is against that, because the latest word is that there is not enough matter in the universe to halt the expansion and bring the universe together again. Furthermore each time the universe collapses, it melts down the world that existed, and if it does emerge from the collapse, an entirely new world results. This is so far beyond our experience in religion or in science that it seems to me it is not worth talking about. Let's talk about the universe we have, which is the one you and I are in. That universe began in a cosmic explosion fifteen billion years ago, and the question arises as to what forces brought about that event. The reply of the cosmologist is that the circumstances of that explosion make it impossible to answer that question by scientific methods.

So, one of the most important questions in the history of human thought, namely: Why do we exist?, or How did we get here?, turns out to have an answer that is beyond the reach of scientific inquiry. That is the point that seems to me to be interesting.

2 What are your views on the conception of an Inflationary Universe?

An Inflationary Universe which springs out of nothing like a bubble seems to me to be so artificial a construct of the theorist's imagination that I find it not to my taste at all. Theoretical physicists are very fond of this kind of speculation. They let their mathematics run wild but may be anchored to very little in the way of observation.

We have one piece of evidence about what the universe was like when it was roughly a million years old, and that is the existence of the primordial fireball radiation and its spectrum. We have one isolated datum that yields information about the universe when it was three minutes old, and that is the ratio of primordial hydrogen to helium. We have no information whatsoever about what happened in the universe when it was younger than three minutes, and in particular, when it was 10^{-43}

second old, and so on. It seems to me naive to construct elaborate theories that propose to answer profound philosophical and religious as well as scientific questions, on the basis of speculation about an area never touched, directly or indirectly, by observation.

3 What do you think of the apparent conclusion of the Stephen Hawking-Roger Penrose paper "The Singularities of Gravitational Collapse and Cosmology" (*Proceedings of the Royal Society of London*, 1970) that space and time had an origin?

What does that mean? That there was a state in which there was no universe? As a point of language, you can say, as Augustine did, that time began when God created the world. So, if you wish, you can use that kind of language. But what does it *mean*? That phrase you used calls for more explanation. Most people want to know the answer to the question of origins, of the First Cause. It's a deeply felt need.

4 On the level of common sense, the layperson implicitly takes it for granted that something cannot come from nothing.

I think he is right. I'll take the ordinary man on the street and his common sense wisdom over the theorists.

5 Do you think the origin of life can be explained purely in terms of physics and chemistry?

What commands my attention is the uncertainty regarding the *probability* that this would happen. It is clear that it has happened once, but that it may happen or has happened more than once is unclear. In fact, scientific knowledge of the probability of life's origin may tell us that we are alone in the universe. Everything I have studied that relates to the properties of the atmosphere of the young earth, the conditions on the young earth, the Urey-Miller experiments, and the like—all that adds up to a plausible explanation as to how life could have sprung from non-living matter. To assume some higher organizing principle outside and above these physical and chemical laws is unnecessary in my view. But that's my opinion. Nobody has demonstrated that life, even a simple bacterium, can evolve from a broth of molecules.

6 How about the theory of evolution as it relates to theism?

Advanced thinkers in theology—I think the majority—take the view that no contradiction exists whatsoever between their faith and the facts of evolution. It is often said that the universe is a continuous act of Creation, that it is a continuously unfolding story whose details could indeed be explained by natural selection. The essence of this thought is that natural selection is the way God works. In other words, there is a process that has some guidance from above, but the details are correctly described by the Darwinists.

7 What in your view would constitute evidence for the existence of a Creator?

This is a problem I have: I am still lacking in religious belief. My view is that of the agnostic. I feel you cannot prove the existence of God, any more than my scientific friends can disprove his existence. I suppose a series of miracles could demonstrate to me that forces are at work in the universe that are utterly outside the reach of human comprehension and that is as close as I can come to the proposition of a proof for the existence of a Creator.

For me, the difficulty is that as a result of my scientific studies, I have a deeply rooted belief in the philosophies of reductionism and materialism—the view that the whole is equal to the sum of the parts—even with respect to questions relating to life and mind.

8 But isn't the perception of God's existence also a matter of personal experience for an individual who is a religious believer?

Perhaps these matters will be illuminated for me at some later time. At the moment—at this stage in my life—I can only perceive that there exist things in the world to which science has no answers, and maybe those things add up to God's existence. There are many faces to reality, and many voices in the universe. Perhaps there are other voices to be heard than those the scientist can hear.

So that is where I am. This is an approach to the existence of God that is very amorphous. That is the direction in which my thoughts are carrying me. But I am not there yet and may never reach that point.

9 Is the Anthropic Principle relevant to the question of God's existence?

On the face of it, this proves the existence of God as well as scientific evidence can. But I believe that there is another explanation for it.

10 In his recent book *A Brief History of Time,* Stephen Hawking states: "If the universe is really self-contained, having no boundary or edge, it would have neither beginning nor end: it would simply be. What place, then, for a creator?" Could you comment on the idea that the universe has no boundary or edge, an idea Hawking describes as speculation?

A universe without a boundary means a "closed" universe—one which oscillates between expansion and contraction, instead of expanding forever. At the end of each contraction stage, the universe is once more in a dense, hot state, in which everything that formed in the previous stage of expansion—every star, planet, and form of life—is melted down and destroyed. In such a universe, there is an infinite number of replays of the Big Bang. It is true, in a formal sense, that in such a universe there is no true beginning—only an infinite repetition of beginnings and endings. But for those who exist in each cycle, there is indeed a beginning—the particular Big Bang that led to their existence. And there is an end—the melting down of all organized forms of matter that precedes the next Big Bang.

I find this all rather formal, as I suggested earlier, and devoid of meaning. One can only say again that all the accepted scientific evidence indicates that about fifteen billion years ago, the universe of which we are now a part was in a very dense, very hot state. It had at that time no stars, no planets, no life—not even any atoms. But the seed of everything that exists today was planted, in a material sense, in that moment. It was literally the moment of Creation. That is as close to scientific evidence for a beginning as one can come. I do not know the meaning of saying, in the light of these circumstances, that there may have been no beginning.

For me, the interesting question is that posed by Leibniz, namely: Why is there something rather than nothing? What forces filled the universe with energy fifteen billion years ago? These are questions of metaphysics—or theology—not physics, but they are very interesting.

8

There Need Be No Ultimate Conflict between Science and Religion

· · · · · · · · · ·

Professor B. D. Josephson

- Born 4 January 1940

- Ph.D. in physics, Cambridge University, 1964; Nobel Prize for Physics (shared with Leo Esaki and Ivar Giaever), 1973; received the Nobel Prize "for his theoretical predictions of the properties of a supercurrent through a tunnel barrier, in particular those phenomena which are generally known as the Josephson effect"

- Currently Professor of Physics, Cambridge University

· · · · · · · · · ·

1 Do you see any conflict between science and religion?

> I don't see a conflict. There are conflicts between the views of many scientists on religion, but I think there need be no ultimate conflict. Science may be capable of extension in a way that is compatible with the tenets of religion.

2 Traditional concepts of a Creator entail such attributes as infinite perfection. Is this incompatible with what we know of the universe?

> I take a slightly modified form of this in the sense that I presume that the Creator acts in as perfect a way as possible, but still there are problems which are unresolvable.

3 On what basis do you believe in such a Creator?

> My experiences in meditation result in such a view.

4 Do you experience such a reality?

> Yes. Rather dimly. But sometimes I experience something that I can be in interaction with, which results in some of the problems and tensions being resolved.

9

The Exquisite Order of the Physical World Calls for the Divine

· · · · · · · · · ·

Professor Vera Kistiakowsky

- Born 9 September 1928

- Ph.D. in physics, University of California, 1952

- Currently Professor of Physics, Massachusetts Institute of Technology

- Areas of specialization and accomplishments: experimental nuclear physics, elementary high energy particle physics, astrophysics; author or co-author of over one hundred papers published in scientific journals and of over one hundred and twenty papers presented at professional meetings; in 1969 she and two friends, Elisabeth Baranger, a theoretical nuclear physicist, and Vera Pless, a mathematician, started a Boston area group called WISE (Women in Science and Engineering); president of the Association of Women in Science in 1982 and 1983

- Professor Kistiakowsky on:

  the origin of the universe: "There remains the question of how the Big Bang was initiated, but it seems unlikely that science will be able to elucidate this. . . . the exquisite order displayed by our scientific understanding of the physical world calls for the divine".

  the origin of life: ". . . the whole process is miraculous, from the formation of the first complex molecules to the evolution of human intelligence".

  the origin of *Homo sapiens:* "There are still gaps to be filled . . . but in principle these can be filled".

  God: "I am satisfied with the existence of an unknowable source of divine order and purpose . . ."

· · · · · · · · · ·

1 What do you think should be the relationship between religion and science?

I think that religion and science are two different approaches to understanding existence. Their domains do not overlap completely. Religion deals with ethical and spiritual matters that are explicitly not the concern of science, and science asks a quantitative understanding of the material world that is irrelevant to religion. The tension between science and religion arises in the regions where they do overlap, three of which are taken up in the next questions. Since religious writings usually date to periods of history when scientific understanding of these questions was non-existent or much less extensive than it is now, there is a clash between the religious portrayal of the particular origin and the current scientific knowledge on that question. I think that it should be acknowledged that religious writing, however divinely inspired, was the product of a human living in a particular period with a particular understanding of the world. In the same way a contemporary scientific result is the product of a human attempt to understand the world, guided by certain criteria of objectivity, but still limited by the training of the mind that produces it. With this understanding, there is no conflict between religion and science. I would not presume to say what their relationship should be for others, but for me they are complementary, dealing with different spheres.

2 What is your view on the origin of the universe: both on a scientific level and—if you see the need—on a metaphysical level?

Cosmology, astrophysics, and the theory of particles and fields have in recent years combined to give us a very coherent theory for the origin of the universe. There are still problems and unresolved questions; astrophysics data are not easy to obtain and calculations are not always possible. If accepted as a still evolving understanding, I am comfortable with our current scientific account. There remains the question of how the Big Bang was initiated, but it seems unlikely that science will be able to elucidate this. The question which is not in the realm of science is, For what purpose? and it is with this that religion can deal uniquely. It is a question that many, including many scientists, consider unnecessary. However, there does seem to be a human drive to find a purpose in existence and thus invoke the will of God in Creation. My acculturation as a scientist makes me uncomfortable with this, but the exquisite order displayed by our scientific understanding of the physical world calls for the divine.

3 What is your view on the origin of life: both on a scientific level and—if you see the need—on a metaphysical level?

I am not well-educated on this topic scientifically, but my understanding is that although research has been closing the gap between the animate and the inanimate, a gap remains. I would not be surprised if this connection were also made. For me, the whole process is miraculous, from the formation of the first complex molecules to the evolution of human intelligence.

4 What is your view on the origin of *Homo sapiens?*

There are still gaps to be filled with respect to the scientific understanding of the details of the evolution of man, but in principle these can filled. The question of why remains the province of religion.

5 How should science—and the scientist—approach origin questions, specifically the origin of the universe and the origin of life?

A clear distinction should be made between the scientific (how) questions and the religious (why) questions, and scientists in their professional capacity should restrict themselves to the former. Their opinions on the latter should not be given especial weight because of their profession but should be respected in direct proportion to the philosophical and theological profundity of the opinions.

6 Many prominent scientists—including Darwin, Einstein, and Planck—have considered the concept of God very seriously. What are your thoughts on the concept of God and on the existence of God?

I believe that God exists but do not have a personal understanding of what this means. The picture held by others of a Being which exists in some real area of the universe I find at best unnecessary. I am satisfied with the existence of an unknowable source of divine order and purpose and do not find this in conflict with being a practicing Christian. The biblical account and the details of church services represent attempts by other humans to deal with the unknowable, and I respect them as such.

10
Was It Planned, Is It Part of a Grander Scheme of Things?

• • • • • • • • • •

Professor William A. Little

• Born 17 November 1930

• Ph.D. in physics, Rhodes, 1955

• Currently Professor of Physics, Stanford University

• Areas of specialization: organic fluorescence; magnetic resonance; low temperature physics; superconductivity; phase transition; chemical physics

• Professor Little on:

the origin of the universe: "I go along with the Big Bang picture but recognize that it does not address the deeper issue as to why it happened—was it planned, is it part of a grander scheme of things?"

the origin of life: "It is hard to believe that all this just happened as a result of the initial conditions. In fact, if that is how it happened, it is all the more remarkable!"

the origin of *Homo sapiens*: "I go along with evolution!"

God: "I . . . might go along with some form of 'intelligence' associated with matter, energy, and the universe".

• • • • • • • • • •

1 What do you think should be the relationship between religion and science?

Religion, as I see it in everyday life, seems more to be concerned with social issues than with the fundamental questions of God and Creation. I think that many scientists, and physicists in particular, are closer to these latter, deeper questions than those in religious studies, as a result of our study of astrophysics, quantum mechanics, relativity, and so on.

2 What is your view on the origin of the universe: both on a scientific level and—if you see the need—on a metaphysical level?

I go along with the Big Bang picture but recognize that it does not address the deeper issue as to why it happened—was it planned, is it part of a grander scheme of things? It is an impressive accomplishment to have all we see about us result from a few laws of physics, lots of energy, and a great deal of patience! It is hard to believe that there isn't more to it than this.

3 What is your view on the origin of life: both on a scientific level and—if you see the need—on a metaphysical level?

The origin of life gets me into the same kind of trouble. It is hard to believe that all this just happened as a result of the initial conditions. In fact, if that is how it happened, it is all the more remarkable! What seems to be missing in the scientific models as they stand today is the absence of any explanation of a driving force or "urge to survive" which I think many of us feel exists in some sense and which could probably be defined in some mathematical sense.

4 What is your view on the origin of *Homo sapiens*?

This bothers me less than the others. I go along with evolution! The way DNA, protein, and the various enzymes work even in the simplest creatures and plants is impressive enough to me, that the antics of *Homo sapiens* are less remarkable.

5 How should science—and the scientist—approach origin questions, specifically the origin of the universe and the origin of life?

I like the "scientific" approach which continues to ask probing and testable questions. In this way it has been possible to push the need for a vital force further and further into the background and to define better and better what still remains of the need. If we do not do it this way it becomes metaphysical and one cannot get an answer of a physical kind.

6 Many prominent scientists—including Darwin, Einstein, and Planck—have considered the concept of God very seriously. What are your thoughts on the concept of God and on the existence of God?

Many prominent scientists have considered the concept of God very seriously; I am sure this is true. But I think the views they have had of God have generally been rather different from those of the clergy. I do not think of God as an old man with a white beard, but might go along with some form of "intelligence" associated with matter, energy, and the universe. I guess I would be

pleasantly intrigued, rather than hostile, and not totally surprised, by a discovery that unveiled a deeper meaning to our
universe, indicative of some such underlying intelligence. I am, of
course, thinking of a piece of *physical evidence*—a leopard does not
change his spots.

11

The Laws of Nature Are Created by God

• • • • • • • • • •

Professor Henry Margenau

- Born 30 April 1901

- Ph.D. in physics, Yale University, 1929

- Currently Emeritus Eugene Higgins Professor of Physics and Natural Philosophy, Yale University

- Accomplishments: President of the American Association for the Philosophy of Science, 1954–1964; former editor of *Journal of the Philosophy of Science, American Journal of Science, Review of Modern Physics, Main Currents of Modern Thought,* and other journals; works include *Foundations of Physics* (with R.B. Lindsay), 1943; *Physics: Principles and Applications,* 1949; *The Nature of Physical Reality,* 1950; *Integrative Principles of Modern Thought,* 1972

- Professor Margenau on:

  the origin of the universe: "God created the universe out of nothing in an act which also brought time into existence".

  the origin of life: "The occurrence of life arose in perfect accordance with the laws of nature".

  the origin of *Homo sapiens:* ". . . *Homo sapiens* is physically an evolutionary follower of the two-legged ape. . . . But God endowed man with a *soul*".

• • • • • • • • • •

There exists a widespread view that regards science and religion in general as incompatible. Let me therefore point out, first of all, that this belief may have been true half a century ago but has now lost its validity as may be seen by any one who reads the philosophical writings of the most distinguished and creative physicists of the last five decades. I am referring here to men like Einstein, Bohr, Heisenberg, Schrödinger, Dirac, Wigner, and many others.

Theories like the Big Bang, black holes, quantum theory, relativity, and the Anthropic Principle have introduced science to a world of awe and mystery that is not far removed from the ultimate mystery that drives the religious impulse. These twentieth-century trends seem to call for a new metaphor in describing the relationship of science and religion.

Nowhere is the tension between science and religion more pronounced than in the origin issues: the origin of the universe, the origin of life, and the origin of *Homo sapiens*. As a scientist and a philosopher of science for over forty years, I have reflected on these questions in my books *Foundations of Physics, The Nature of Physical Reality,* and *The Miracle of Existence.*

These issues have now drawn me to an even more extensive exploration. I would like to map modern scientific perspectives on these issues. To this end, I have compiled views on the three main origin issues from some of the most noted scientists of the present day, and these views are presented in this anthology. I have below my responses to the six questions which form the fundamental framework of *Cosmos, Bios, Theos*:

1 What do you think should be the relationship between religion and science?

The first question concerns the relation between science and religion. It has been of interest to the editors of this book for many years. I have written two books, one of them dealing with the epistemology of science,[1] the other relating it to religion.[2] I will therefore take the liberty to present my view here in personal terms. A book entitled *Science and Religion* by Harold Schilling[3] is also recommended, for it contains a large number of factual instances allowing comparisons between the two fields.

Science deals with the laws of nature as they are understood by man. They are based on observations made by man including experiments, that is, experiments arranged or made possible by human intervention. In addition to them the laws are designed to account for phenomena independent of human preparation, such as occur in astronomy, many parts of physics and chemistry, and, perhaps with restricted frequency, in biological sciences, which include medicine.

There is a simple scheme which formulates the method by which science advances. It will here be briefly reviewed and illustrated. In Figure 1, P represents the "plane of perceptions", to the left of it lies the C-field of conscious mental responses to the perceptions. The P-plane extends in and out of the paper, and the C-field is three-dimensional. The letter C stands for "construct", not "concept", a term which designates a set of related constructs.

Double lines connecting specific observations, that is, measurements (points on the P-plane) with constructs, represent what in my earlier writings I called "rules of correspondence". They are identical with the term "operational definition" in the language of the famous Professor Bridgman, who was a friend of mine. Finally, I note that the formulation of laws of nature involves first the combination of constructs into concepts which are combinations of measurements, and then the application of mathematics.

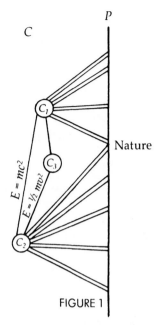

FIGURE 1

The theory of knowledge, the epistemology I have formulated here, applies to physics, chemistry, biology, and their combinations, but not to psychology and the arts. It does lead to all the laws of nature but does not account for their origin. They surely could not have developed by chance or accident. What, then, is the answer to the question concerning the origin of the innumerable laws of nature? I know only one answer that is adequate to their universal validity: they were created by God.

The preceding epistemological considerations led me to a study of the physical processes performed by the human brain, which are most clearly discussed in the works of Sir John Eccles, the famous neurophysiologist. Many of the entities involved are of atomic and molecular size and mass, and their behavior cannot be described exclusively by classical physics: it requires the use of quantum mechanics. This means that the description of a brain state, which has not yet been achieved as a totality, will appear in the form

$$\Psi = \Psi_1 (x_1) \, \Psi_2 (x_2) \, \Psi_3 (x_3) \ldots \Psi_n (x_n)$$

where each is a Schrödinger-type state function satisfying the rules of quantum mechanics. This requires that

$$|\Psi|^2 \, dx_1 \, dx_2 \ldots dx_n = 1$$

Each $|\Psi_1|^2$ defines a probability, not a classical state for which the coordinates are known precisely. Hence the total state of the brain at any time defines a probability (not a classical state) for which the coordinates are known precisely. Hence the total state of the brain at any time defines a probability, not a fixed coordination of individual values x_1.

In ordinary terms this means that if the brain determines the mind, it too will be at any time in a state of probability; physics alone will not determine its behavior. From its physical condition we cannot predict the action which it induces the body to perform; we are forced to assign to it the power to select from actions it is capable of performing for physical reasons: we conclude that the mind can choose actions which physics permits. In other words: there is freedom of the will. The mind is not a mere mechanism but has a choice. Quantum mechanics accounts for freedom of the will.

The preceding discussion of the interaction between the human mind and the brain leads us to comment briefly on the mind-body problem in general. The enormous success which has been achieved in the construction of mechanical computers has convinced many scientists and others that a perfect mechanical computer, a robot, could be identified with our mind. In other words, our mind is identified as a highly complex and perfect mechanical device, a robot, which acts in complete accordance with the laws of nature. This view omits, of course, what was presented in the foregoing section concerning the actions of the human brain and mind. It converts psychology into a branch of physics and chemistry.

This conclusion, however, encounters a grave difficulty at the very beginning of the effort to establish it. Physics and chemistry deal exclusively with fundamental quantities subject to operational definitions: force equals mass times acceleration, energy equals Planck's constant times frequency of light, and so forth. But how do we define, or measure, mental states? We speak of brilliant joy, a dark mood, deep sorrow, but know of no quantitative way of measuring such states. This is generally true of the laws through which the states of our mind are connected. Psychology is devoid of operational definitions, although attempts are sometimes made to establish them.

Several decades ago, a friend of mine, a distinguished psychologist, performed an experiment in which he attempted to establish mathematical laws of psychology in terms of measurable and meaningful quantities. He designed an apparatus which exerted a measurable force on a patient's arm. As the force increased, the pressure on the arm increased, and the patient reported when he felt the change in pressure. This began to establish a relation between pressure and intensity of sensation. It turned out that at the beginning of the experiment the pressure felt was proportional to the logarithm of the stimulating force, but as the pressure increased and resulted in the beginning of pain in the patient's arm, the law broke down.

The language used to describe mental states, which is devoid of operational definitions, is thus forced to use metaphors. We speak of "brilliant joy", "a dark mood", "deep thought", and many similar non-operational combinations. What all this means is that the mind cannot be treated as a body; it is a unique entity which affects and regulates the human body. Animals, too, have minds which affect their behavior; they are similar to ours but different in one fundamental way: they lack ethics and religion, and their freedom of the will is limited by physical circumstances.

The relationship between religion and science must in my opinion be further described as follows: every science faces as an abstract foundation two problems—the discovery of the *laws of nature* and their application for the maintenance and the welfare of mankind. The definition of a law seems clear, but I should add here that by nature I refer to the total universe, not merely to the natural phenomenon Earth.

In order to be accepted by scientists a law of nature must possess three important properties: *simplicity, extensibility* (that is, applicability to all similar observable phenomena), and a kind of mathematical and philosophical *elegance*. This third property is sought, but not always obtained. For instance, the force between two spheres of mass m was found by Newton to be proportional to m^2/r^2, r being the distance between their centers. Recently a physicist claimed to have discovered that the exponent 2 in the denominator of this expression differed slightly from 2. Even now, every physicist seems disappointed by this result and efforts are being made to prove it false.

Now arises the question: What is the origin of the laws of nature? For this I can find only one convincing answer: they are created by God, and God is omnipotent and omniscient. In my latest book *(The Miracle of Existence)* I called Him "the Universal Mind" and suggested that every human soul is a part of Him. This conception of God is not to be confused with some crude version of pantheism (since pantheism identifies the world with

God and is therefore tantamount to atheism): every human soul is a part of God in the highly specialized sense of modern quantum physics and in the sense understood by the great mystics like St. Paul, who talked of the Ultimate Reality "in whom we move and live and have our being".

Hence my answer to Question 1: Science needs religion in order to account for its origin and its successes. I discussed this view with Einstein when I did research at the Institute for Advanced Study at Princeton in 1932 and remember his comment: "The discovery of a fundamental, verified law of nature is an inspiration of God".

There is a common belief that science rejects miracles. But what, in precise terms, is a miracle? I suppose the answer has to reflect the fact that science is not yet, and probably will never be, a complete system of explanation. It is important to keep this reservation in the background of our considerations. The existence of man, indeed of the entire universe, has long been regarded as a miracle, incomprehensible without assuming the existence of a Creator who is omnipotent and omniscient. But did it violate the laws of nature? During this century the Big Bang theory was formulated and confirmed. A very small but extremely massive sphere of matter, in many respects similar to a black hole, could spring out of nothing without violating any known law of nature. To be sure, it would be most unsatisfactory to regard this as an accident. God, however, created not only the physical universe but also the laws which it has to obey. This latter fact is often ignored, not only by certain biologists but by theologians as well. Schleiermacher, to be sure, was an exception. For he said in his *Speeches to the German Nation:* the existence of the laws of nature is the greatest of all miracles.

2 What is your view on the origin of the universe: both on a scientific level and—if you see the need—on a metaphysical level?

God created the universe out of nothing in an act which also brought time into existence. Recent discoveries, such as observations supporting the Big Bang and similar astronomical phenomena, are wholly compatible with this view.

3 What is your view on the origin of life: both on a scientific level and—if you see the need—on a metaphysical level?

The occurrence of life arose in perfect accordance with the laws of nature. Plant life and, in fact, some elementary forms of animal life developed in harmony with Darwin's theory of evolution, which does not contradict Christian and other religious doctrines. There is one aspect, however, on which I would like to elaborate briefly.

Recently my daily newspaper, the *New Haven Register,* cited a book titled *The Mystery of Life's Origin* written by three scientists. It said: "They explain that life would not have started by chance and argue that a Creator beyond the cosmos is the most plausible explanation for life's origin". Supporting this view the article mentions that the British astronomer Fred Hoyle is widely noted for the statement that believing the first cell originated by chance is like believing a tornado ripping through a junk yard full of airplane parts could produce a Boeing 747. Other similar examples follow.

4 What is your view on the origin of *Homo sapiens*?

I believe that *Homo sapiens* is physically an evolutionary follower of the two-legged ape. But here I must add a highly important but usually ignored distinction: apes may have minds, to be sure of simple types and not quite comparable with the brain and the mind of man. But God endowed man with a *soul,* an abstract entity capable of making contact with Him. Here I am arriving at a point which recalls the view presented in my recent book, *The Miracle of Existence.*

5 How should science—and the scientist—approach origin questions, specifically the origin of the universe and the origin of life?

6 Many prominent scientists—including Darwin, Einstein, and Planck—have considered the concept of God very seriously. What are your thoughts on the concept of God and on the existence of God?

These questions have already been answered.

• • • • • • • • • •

Notes

1. *The Nature of Physical Reality* (Woodbridge, CT: Oxbow Press, 1977)
2. *The Miracle of Existence* (Woodbridge, CT: Oxbow Press, 1984)
3. *Science and Religion* (New York: Scribner and Sons, 1962)
4. The importance of simplicity became clear to me after I published one of my earliest books, entitled *The Mathematics of Physics and Chemistry,* of which my friend George Murphy, the chemist, was the co-author. It was praised highly by reviewers for the simplicity of its contents and sold exceptionally well.

12

Science Will Never Give Us the Answers to All Our Questions

• • • • • • • • • •

Professor Sir Nevill Mott

• Born 30 September 1905

• M.S. in theoretical physics, Cambridge University, 1930; Nobel Prize for Physics (shared with Philip W. Anderson and John H. Van Vleck), 1977; received the Nobel Prize with Anderson and Van Vleck "for their fundamental theoretical investigations of the electronic structure of magnetic and disordered systems"

• Cavendish Professor of Physics at Cambridge University until 1971

• Works include *Atomic Structure and the Strength of Metals,* 1956; *Electronic Processes in Non-Crystalline Materials* (with E. A. Davis), 1971; *Elementary Quantum Mechanics,* 1972

• Professor Mott on:

the origin of the universe: ". . . the concept of the Big Bang and what followed it are strongly supported by observations".

the origin of life: ". . . it arose through some chemical reaction in the primeval mud".

the origin of *Homo sapiens:* "Whether a specific mutation separates *Homo sapiens* from the other animals I do not know".

God: ". . . we can and must ask God which way we ought to go, what we ought to do, how we ought to behave".

• • • • • • • • • •

Many people of my generation came from families with some religious beliefs, which they abandoned when they grew up. This was so also of my parents, and I was brought up to respect Christian ethics but not to take part in any Christian worship. At about the age of fifty, influenced by various people, I started attending church, found that I could express in that way a vague belief in God, and ever since have been asking myself how much of the established doctrine of the Anglican church I could believe. I wanted to understand the relation between scientific and religious truths. I knew, of course, that all scientific theories are provisional and may be changed, but that, on the whole, they are accepted from Washington to Moscow because of their practical success. Where religion has opposed the findings of science, it has almost always had to retreat. I thought then, and still do, that science can have a purifying effect on religion, freeing it from beliefs from a pre-scientific age and helping us to a truer conception of God. At the same time, I am far from believing that science will ever give us the answers to all our questions.

Religious doctrines, as stated in church services, seem to me beliefs that have been held by our ancestors and are therefore worthy of respect, and about which we should meditate and see if they can help us now in our understanding of God. They are certainly not accepted from Washington to Moscow. We can, I believe, at least within the reformed churches, accept or reject parts of them, or interpret them in terms of the thought of the twentieth century. Even within the Catholic church, too, some prominent theologians have felt free to interpret them in this way.

This article is about origins, and historically our views on the origin of the universe have been profoundly influenced by religion. At present the concept of the Big Bang and what followed it are strongly supported by observations, and I have no reason to doubt current theories. To me the most fascinating part of them is the Anthropic Principle, the undoubted fact that the constants of nature, for instance the number of particles in the universe or the ratio of the electron's mass to that of the proton, have values just right to allow the primeval gas to condense into nebulae and stars and occasionally form planets on which we can live. Were they only slightly different, this could not have happened and we could not exist. In the future, perhaps, a mathematical theory will show that these constants could not be otherwise than they are—in which case an anthropic mathematical theory will be a very curious conclusion to our endeavors. Or perhaps it was all planned by a high intelligence, a God, who did it with us in mind. I do not know and do not expect to know, neither do I very much care. If it was planned by God, he is so far away from the ever-present God of worship and prayer that I find it hard to imagine that they are the same being.

As regards the origin of life, I can only follow modern evolutionary theory and believe that it arose through some chemical reaction in the primeval mud. Whether a specific mutation separates *Homo sapiens* from the other animals I do not know. It is certainly possible. If not—and if we hold to the doctrine of eternal life in any literal sense—we have to face some harsh questions: do our cats and dogs, and wild animals, have some form of eternal life too?

Let me now face Professor Margenau's last, and to me the most important, question: What is my concept of God? I need hardly say that this is entirely personal; I most certainly do not expect or wish to carry conviction with those who think otherwise.

In my understanding of God I start with certain firm beliefs. One is that the laws of nature are not broken. We do not, of course, know all these laws yet, but I believe that such laws exist. I do not, therefore, believe in the literal truth of some miracles which are featured in the Christian Scriptures, such as the Virgin Birth or water into wine. This is not consciously because I am a scientist; if God is all-powerful, of course he could break his own laws. My disbelief is because I am repelled by the idea of a God who would perform miracles of that kind to mark a special occasion in the history of his revelation of his nature to us. Miracles, also, tempt us to ask, why does not God perform more of them? Why did he not stop the Black Death or an earthquake in Armenia? God works, I believe, within natural laws, and, according to natural laws, these things happen.

In considering God's power, we must not look for a God of the Gaps, a god who is called in for those phenomena for which there is yet no scientific explanation. But I believe that there is one "gap" for which there will never be a scientific explanation, and that is man's consciousness. No scientist in the future, equipped with a super-computer of the twenty-first century or beyond, will be able to set it to work and show that he is thinking about it. This has been argued by my successor as the head of the Cavendish Laboratory at Cambridge, Sir Brian Pippard, in an essay entitled "The Invincible Ignorance of Science" (*Contemporary Physics*, vol. 29 [London: Taylor and Francis, 1988], p. 393). Pippard, an agnostic about God, does not describe this as the "gap" where God makes himself known. But I would deduce from this hypothesis that the way God plays a part in our lives is because countless men and women claim to be conscious of him, when they seek him, and accept that he is the God of love. God can speak to us and show us how we have to live.

God as we know him, it seems to me, cannot desire the earthquakes, famines, and plagues that haunt mankind. Or perhaps we should say that he has voluntarily relinquished his omnipotence—a concept which may bring these statements into line with our prayers in church to the Almighty and with belief in a God who engineered the Big Bang. And I do not think that we can consider this God with whom we live as omniscient, either. This is basically because I believe in free will, as Dr. Johnson did when he said, "Sir, we know the will is free and there's an end on it". So it seems to me that what we do is not predetermined and that this means that God cannot know the future. The lines from a Moslem prayer (from Raman Baba, who lived in what is now Afghanistan in the seventeenth century, quoted in *The Oxford Book of Prayer*, Oxford University Press, 1985; p. 339):

> All the pages not yet written He has read,
> Perfect knowledge of all secrets has my Lord

are beautiful, but not, I believe, true. If we wish to believe in free will and in God's omniscience, I believe we must think of God as "outside time", which is a difficult concept for most of us.

Turning now to physics, I follow the Copenhagen Interpretation of quantum mechanics. I learned it from Niels Bohr, whose pupil I was for a period, and I have used it in all my research over sixty years. I believe therefore that chance is a real element in nature. Thus, out of billions of atoms in a radioactive specimen, we know that a given number, more or less, will break up today, but we do not know which ones. The Uncertainty Principle tells us that if we tried to find out which ones were, so to speak, getting ready to break up, we would have to probe the nuclei, and this might well spark off the disintegration itself. Chance, then, is a real feature of our universe, though many would deny this. Chance, I believe, is not just a cover-up for our ignorance. In theological terms, I must maintain that God himself does not know which atom will break up today and which next year. I do not claim, however, any relationship between the Uncertainty Principle and human free will. But the Uncertainty Principle does teach me that nature is not deterministic in the Newtonian sense and that there is nothing in physics against my belief in free will. And so, turning to religious history again, it was not predicted that Pontius Pilate would behave as he did. I conclude that the great drama of the New Testament might not have happened.

Here then is a description of God. He is a being who does not break natural laws or know the future, but who can help us to choose the way we live, if we ask him, and so profoundly affect what happens in his world. The miracles of human history are those in which God has spoken to men. The supreme miracle for Christians is the Resurrection. Something happened to those few men who know Jesus, which led them to believe that Jesus yet lived, with such intensity and conviction that this belief remains the basis of the Christian Church two thousand years later. Was it a "bodily" resurrection? I will quote Hans Küng, the Catholic theologian at the University of Tübingen, whose teaching did not always please the Vatican. He writes (in *On Being a Christian* [London: Collins, 1974], p. 361), "Anyone who perceives the real point of the resurrection message will regard some fiercely contested [historical] questions as peripheral". Not, if I understand his views, as simply a movement of molecules. I am impressed too by the point of view of the present Archbishop of York (Dr. John Habgood, *Science and Religion*, [London: Hodder and Stoughton, 1964]), that to understand the Bible we must try to enter into the belief patterns of the period—a hard saying, perhaps, as it may shut off the poorly educated from a full understanding of Christianity.

I believe, then, that we can and must ask God which way we ought to go, what we ought to do, how we ought to behave. And in doing this, we must not be too impressed by the insights of the end of the twentieth century in molecular biology and modern astrophysics. It has been said that he who tries to marry his religion to the beliefs of one period risks widowhood in the next. We must respect the wisdom and insights of our ancestors. This is why, for me, it is possible to worship in church and recite a creed, much of which I do not believe.

At the same time, I realize that the answers God gives to those who ask him are not always the same. In religion I am proud to call myself a liberal, but I feel that in the last thirty years liberalism has become a dirty word. And, alas, in history we see that the answers given to those that asked him may shock us now, whether we are liberals or not. Christians have asked, and the answer has been that crusading armies should storm Jerusalem and kill all Jews and Moslems there, and later that heresy is so dangerous for our salvation that heretics must be burned. In another great religion, the answer has been given quite recently that an apostate must be killed. Can we not believe that our understanding of God does greatly change with time and place? Talking to a Jewish friend, I asked him how he reconciled his conception of God with that described in the thirty-first chapter of the Book of Numbers, where Jehovah commands the Israelites to kill all the men of the Midianites but to keep the girls for themselves. "Ah", he said, "God has grown up since those days". Perhaps—but not, I fear, everywhere. Nonetheless, what we hear when we ask him is, I believe, the only way to meet this God who lives amongst us. What we believe about him can affect our daily life, our politics, our educational system, and how we think about any differing role of women and of men in our society. But, the question asked in this article is, can we relate this God to the Creator of the universe?

Scientists, notably Guth and Vilenk in the U.S., Linde in the U.S.S.R., and Hawking in England, are pushing the story of the Creation back before the Big Bang, to a universe empty but full of energy, subject to fluctuations, one of which triggered off the explosion. And as for the Anthropic Principle, as I have said, others like to speculate that there may have been—or are—many other universes, and it was just chance that one of them had the constants just right for us. I would rather believe that the laws of nature will be found to be such that they could not be otherwise, and that the properties that also allow us to exist follow from the ultimate mathematical equations. When, or if, our understanding gets to this point, as Stephen Hawking puts it, "physics will be finished"—his kind of physics, anyhow.

But however we look at it, here is our universe and we ask: Must it have a Creator and Intelligence behind it? This is a question to which perhaps we shall never have an answer but never stop asking. This all-powerful God—perhaps outside time, who set the universe going, looked ahead to man, saw that the constants of nature were right, sat through the millennia while evolution and the struggle for existence produced man, and thought about him from the beginning—is one which I find hard to reconcile with the God of Love, who, Christians believe, is among us now. It is wise, I believe, not to worry too much about questions that cannot be answered or seek to find in the Anthropic Principle evidence for the God who matters to us.

At the same time, I cannot forget the old Jewish story, of how God talked to Abraham, and said, "But for me, you would not be here". "I know that, Lord", answered Abraham, "but were I not here there would be no one to think about you". We shall go on doing it.

13

Religion and Science Both Proceed from Acts of Faith

• • • • • • • • • •

Professor Robert A. Naumann

- Born 7 June 1929

- Ph.D. in physical chemistry, Princeton University, 1953; received the Alexander von Humboldt Stiftung Senior U.S. Scientist Award in 1983

- Currently Professor of Chemistry and Physics, Princeton University

- Areas of specialization and accomplishments: radioactivity; inorganic chemistry; nuclear physics; conducting research for the National Science Foundation in the areas of electromagnetic isotope separation, radiochemical separation procedures, the examination of nuclear structure through radioactive and charged particle nuclear spectroscopy, the implantation of radioactive isotopes into solids, and the formation and properties of muonic atoms

- Professor Naumann on:

    the origin of the universe: ". . . current questions now arising in cosmology, elementary particle physics, and microbiology have an obvious metaphysical or religious content".

    the origin of life: ". . . given the right precursors and conditions, living systems will arise spontaneously. Nevertheless, we will continue to be confronted by the deeper questions: What are the purposes and consequences of living beings in the universe?"

    the origin of *Homo sapiens:* "I am content with a terrestrial evolutionary sequence beginning with amino acids and leading to *Homo sapiens*".

    God: "The existence of the universe requires me to conclude that God exists".

• • • • • • • • • •

1 What do you think should be the relationship between religion and science?

2 What is your view on the origin of the universe: both on a scientific level and—if you see the need—on a metaphysical level?

> I find religion and science present no conflict. Both proceed from acts of faith. The religious follower holds that the universe evolves according to God's plan. The scientists believes that a few fundamental and potentially comprehensible principles can explain the mechanism he observes in the universe. I firmly agree that current questions now arising in cosmology, elementary particle physics, and microbiology have an obvious metaphysical or religious content.

3 What is your view on the origin of life: both on a scientific level and—if you see the need—on a metaphysical level?

> We now appear to know a great deal about the mechanisms operating in living plants and animals. It may well be possible within the next decades to assemble, in the laboratory, a self-replicating virus starting from inert simple organic chemicals. If so, we should assume that, given the right precursors and conditions, living systems will arise spontaneously. Nevertheless, we will continue to be confronted by the deeper questions: What are the purposes and consequences of living beings in the universe?

4 What is your view on the origin of *Homo sapiens*?

> I am content with a terrestrial evolutionary sequence beginning with amino acids and leading to *Homo sapiens*. It strikes me that we humans display great hubris in constructing a life picture with *H. sapiens* at the evolutionary pinnacle. The discovery of other and possibly superior life-forms in the universe would cause a profound and perhaps needed re-evaluation of our niche and relative importance in the universe.

5 How should science—and the scientist—approach origin questions, specifically the origin of the universe and the origin of life?

> By being open-minded, objective, and doing the best science of which we are capable.

6 Many prominent scientists—including Darwin, Einstein, and Planck—have considered the concept of God very seriously. What are your thoughts on the concept of God and on the existence of God?

I hold that God is the totality of the universe; this includes all scientific principles, all matter and energy, and all life-forms. The existence of the universe requires me to conclude that God exists.

14

Our Final Ineptitude at Producing a Rational Explanation of the Universe

Professor Louis Neel

• • • • • • • • • •

- Born 22 November 1904

- Ph.D. in physics, University of Strasbourg, 1932; Nobel Prize for Physics (shared with Hannes Alfven), 1970; received the Nobel Prize "for fundamental work and discoveries concerning antiferromagnetism and ferromagnetism, which have led to important applications in solid-state physics": works include *The Selected Works of Louis Neel* (with Nicholas Kurti), 1983

- Director of the Center for Nuclear Studies, Grenoble, since 1956

• • • • • • • • • •

1 What do you think should be the relationship between religion and science?

Religion and science are two very separate domains. Any attempt to merge them can only distort them without any advantage. The progress of science, no matter how marvelous it appears to be, does not bring science closer to religion but it leads to dead ends and shows our final ineptitude at producing a rational explanation of the universe.

2 What is your view on the origin of the universe: both on a scientific level and—if you see the need—on a metaphysical level?

As a physicist, I consider physics to be an experimental science. A hypothesis is of interest only if it is possible to verify its consequences by discovering new phenomena or new directions. This means that all hypotheses concerning the origin of the universe do not belong to physics but to metaphysics or to philosophy and that physicists as such are not qualified to deal with them.

3 What is your view on the origin of life: both on a scientific level and—if you see the need—on a metaphysical level?

4 What is your view on the origin of *Homo sapiens*?

5 How should science—and the scientist—approach origin questions, specifically the origin of the universe and the origin of life?

> Personally I am interested only in problems that I have some hope to solve. Man is quite conceited to speculate on the origin of life and on his own origin.
>
> If everything is limited to the material environment, one cannot conceive how the brain, a minute part of an immense universe in time and space, could understand it: one might as well ask the piston of an automobile motor to retrace the history of the automobile. If a God exists outside of our material universe, it is just as ridiculous to hope to penetrate his purposes.

6 Many prominent scientists—including Darwin, Einstein, and Planck—have considered the concept of God very seriously. What are your thoughts on the concept of God and on the existence of God?

> To give my opinion on God's existence, it would be necessary to define exactly the concept; unfortunately, for three thousand years that concept has taken infinite forms.

> **Personal addendum:** I advise scientists and philosophers to read or read over again Ecclesiastes as well as St. Paul in his First Letter to the Corinthians xiii: 1–13 with great humility.

15

A Feeling of Great Surprise That There Is Anything

· · · · · · · · · ·

Professor Edward Nelson

- Born 4 May 1932

- Ph.D. in mathematics, University of Chicago, 1955

- Currently Professor of Mathematics, Princeton University

- Professor Nelson on:

  the origin of the universe: "Until we have a sound and rigorously established cosmology, it is premature to investigate the origin of the universe".

  the origin of life: "It seems plausible that natural selection played a role even before the emergence of life".

  the origin of *Homo sapiens:* "I subscribe to the view, not prevalent today even among religious people, that something went basically wrong with our origin (original sin)".

  God: "I believe in, pray to, and worship God".

· · · · · · · · · ·

1 What do you think should be the relationship between religion and science?

Scientists choose the problems on which they work, and are sustained by hopes of discovery, on grounds deeper than rational considerations. For some scientists, such choices and hopes are religious. But in the day-to-day work of doing science, it is essential that the scientist simply ignore religious beliefs. This is because science only succeeds within a limited, restricted methodology. An analogy would be the work of an accountant auditing the books of a charitable organization: nothing good would come in being influenced by the worthy aims of the organization when doing the audit.

Although at times, especially in studies of origins, science deals with penultimate questions, its methods are incompatible with those of theology. Theology can ponder with benefit the results of science, but not vice versa. In my view, dialogue

between science and religion as subjects is misconceived, although dialogue between scientists and theologians can perhaps be fruitful.

This is a time when, especially in America, science is under vigorous attack from creationists, radical animal rights activists, and doctrinaire opponents of genetic engineering. We need the help of reasonable religious people in combating these dangerous simplistic assaults, which are sometimes mounted in the name of religion.

2 What is your view on the origin of the universe: both on a scientific level and—if you see the need—on a metaphysical level?

I have no scientific expertise in questions of the origin of the universe, but I am highly skeptical about contemporary cosmology. Scientists are as prone to fads and herd behavior as the rest of our species, and in my view the Big Bang theory has been too widely accepted on the basis of insufficient evidence. I am persuaded of this partly by the work of Irving Segal (*Mathematical Cosmology and Extragalactic Astronomy*, New York: Academic Press, 1976; and many articles). Until we have a sound and rigorously established cosmology, it is premature to investigate the origin of the universe.

On a metaphysical level, I believe that the origin of the universe is to be found in the free act of its Creator.

3 What is your view on the origin of life: both on a scientific level and—if you see the need—on a metaphysical level?

Again, I have no scientific expertise on the origin of life. I find Freeman Dyson's book *The Origins of Life,* with emphasis on the plural, highly intriguing. It seems plausible that natural selection played a role even before the emergence of life. One problem that perhaps mathematicians could fruitfully address is this: What structure is necessary in a dynamical system for the emergence of natural selection? Is it possible in a cellular automaton with simple rules such as John Conway's *Game of Life*?

I do not think that Dyson was wrong (in *Infinite in All Directions*) in elevating maximal diversity to the status of a metaphysical principle. I believe that this is such a deep part of the will of the Creator that it is of great ontological and moral significance.

4 What is your view on the origin of *Homo sapiens*?

I am utterly unqualified to express any scientific views on the origin of *Homo sapiens*.

I subscribe to the view, not prevalent today even among religious people, that something went basically wrong with our origin (original sin). William Golding's novel *The Inheritors* portrays us as we might have been and as we are.

5 How should science—and the scientist—approach origin questions, specifically the origin of the universe and the origin of life?

Scientists should approach origin questions as they approach any other questions, with skeptical audacity, seeking discoveries with falsifiable content.

6 Many prominent scientists—including Darwin, Einstein, and Planck—have considered the concept of God very seriously. What are your thoughts on the concept of God and on the existence of God?

One of my earliest memories is a feeling of great surprise that there is anything. It still strikes me as amazing, and for me this is the fundamental religious emotion. I believe in, pray to, and worship God.

16
Creation Is Supported by All the Data So Far
· · · · · · · · · ·
Dr. Arno Penzias

- Born 26 April 1933

- Ph.D. in physics, Columbia University, 1962; Nobel Prize for Physics (shared with Pyotr Kapitza and Robert W. Wilson), 1978; received the Nobel Prize with Wilson "for their discovery of cosmic microwave background radiation"

- Currently Vice-President for Research, AT&T Bell Laboratories

- Dr. Penzias: ". . . astronomy leads us to a unique event, a universe which was created out of nothing, one with the very delicate balance needed to provide exactly the conditions required to permit life, and one which has an underlying (one might say 'supernatural') plan. Thus, the observations of modern science seem to lead to the same conclusions as centuries-old intuition".

· · · · · · · · · ·

Human wonder at the world around us predates recorded history. When the psalmist said, "What is man that thou art mindful of him?", that was surely not the first time that someone looked up at the heavens and wondered what this world was all about. Archeological evidence—from the crescent moons scratched on the walls of ancient caves to the giant blocks at Stonehenge—testifies to our longstanding concern with cosmic questions.

While wonder necessarily began with individual awareness, the growth of civilization has led to the integration of experience. As our sophistication increases, we habitually categorize the sum of experience, using names like "theology" and "astrophysics". But categorization leads to separation. In particular, our understanding of the world around us has grown along two parallel courses, based on largely separate portions of the entirety of human experience.

One portion encompasses what I might call tangible experience—the kind of knowledge that goes with "one and one is two", "water is wet", and "rocks are heavy". The other portion deals with intangible things like love, faith, humanity, and the feeling of order and purpose in the world. While both kinds of knowledge are valid, one or the other has enjoyed the greater share of popular respect at different times in human history. Our intangible knowledge and our quantitative knowledge seem to conflict from time to time, and seem to force us to choose between them. In no field is that conflict more apparent than in cosmology, in thinking about the world around us, in thinking about the universe.

In ancient times, theology outweighed the barely-formed precursors of physical science. But physical knowledge soon began to grow in prestige as well as size. By the end of the Middle Ages, theology could no longer ignore science. The resulting dichotomy between tangible and intangible knowledge perplexed many of our own great scholars—none greater than the Rambam himself. When Rabbi Moses Ben-Maimon thought about the universe (or, to be more precise, when he thought about thinking about the universe), he advised his readers not to be swayed by empirical data.

> In short, in these questions, do not take notice of the utterances of any person. I told you that the foundation of our faith is the belief that God created the Universe from nothing; that time did not exist previously, but was created; for it depends on the motion of the sphere, and the sphere has been created.[1]

Maimonides's "dogmatic" position that the universe was created out of nothing conflicted with "empirical" data—data from none other than Aristotle himself—that matter was eternal.

How could the everyday person take sides in this dispute between giants? The effort involved in trying to fit dogma and fact into the same mind seems too difficult. We can picture Maimonides's reader—wanting to hold on to the teaching of the faith, but as a rational person wanting to keep a grasp on everyday facts—being pulled by two opposing "truths". One held that the universe was created out of nothing, while the other proclaimed the evident eternity of matter. The "dogma" of creation was thwarted by the "fact" of the eternal nature of matter.

Well, today's dogma holds that matter is eternal. The dogma comes from the intuitive belief of people (including the majority of physicists) who don't want to accept the observational evidence that the universe was created—despite the fact that the creation of the universe is supported by all the observable data astronomy has produced so far. As a result, the people who reject the data can arguably be described as having a "religious" belief that matter must be eternal. These people regard themselves as objective scientists. The term "Big Bang" was coined in a pejorative spirit by one of these scientific opponents who hoped to replace the evolutionary universe idea with a Steady State theory—one which said that the universe has always looked exactly as it looks now. More recently, this now-discredited attempt has been replaced by an Oscillating Universe theory, one in which the cosmos explodes and collapses throughout eternity.

If the universe hadn't always existed, science would be confronted by the need for an explanation of its existence. Since scientists prefer to operate in the belief that the universe must be meaningless—that reality consists of nothing more than the sum of the world's tangible constituents—they cannot confront the idea of creation easily, or take it lightly. Well, I hope that we, as modern people, might be able to leave dogma aside and be willing to look at facts, at least the facts as we understand them today.

Let us use our eyes and our intellect to see what the world is really like. Let us look at the world—out at the heavens around us—not through a telescope (because that would be a little difficult to do in the daytime) but perhaps through a library. I invite you to examine the snapshot provided by half a century's worth of astrophysical data and see what the pieces of the universe actually look like. Here are the galaxies, clouds of stars, each with as many as a hundred billion stars clumped together in what amount to mere specks of dust on a universal scale. When I try to depict the universe for non-technical people, I often ask them to think of a large room like this one, and to imagine all the little invisible particles of dust floating in the air within it. If you could image that each one of the billions of dust particles in this room were itself a galaxy of a hundred billion suns, you would begin to appreciate the vastness of just a small corner of the universe.

The pieces of the universe, the galaxies, are flying apart one from the other. When we look out from our own galaxy, the Milky Way, at distant objects in the sky—galaxies which are so far away that their light takes tens of millions of years just to get here—we find that they are all moving away from us. Furthermore, we find that the further away any one is from our own, the faster it is moving away from us. The relation between distance and speed constitutes the basic observation about the nature of the world we live in. The link between "further" and "faster" is just about all we need to tell the "Big Bang" story.

It is a hard concept to grasp, but I have sometimes been able to explain it to fourth-graders. In such a school atmosphere, I often begin by asking about teacher's pets. "Do your teachers have favorites? Do they treat you all fairly? Suppose you came late to a gym class and a foot race was underway, how would you know whether or not the teacher had gotten the race off to a fair start?" With luck, I get the children to answer, "The fastest child should be out in front; the slowest child should be in the rear". They understand that the faster a child runs, the further that child should be from the starting point at any given moment. And then I ask, "What makes a race fair to all its contestants?" What they tell me is, "All the children have to start from the same place at the same time".

So the observation that the fastest galaxy is the furthest away merely says that the race of the galaxies is a fair one. All the galaxies began their flight one from another, "from the same place at the same time". To build a quantitative picture we must use the laws of physics. As we do our calculations, we can move backwards in time to earlier and earlier epochs at which the pieces were closer and closer to the start of the race. (Moving away at sixty miles an hour, for example, a contestant was one mile from that start after one minute, and just about one inch after one millisecond.)

At the start, each of these pieces needed a staggeringly large amount of energy just to escape from the gravitational pull of its neighbors. Furthermore, for this escape to have been at all possible, the matter from which the pieces were made, as well as this tremendous energy, seem to have appeared out of nothing in an instant. All the energy (as well as any matter present at the event) ought to have appeared out of nothing because, had all this material sat there in a quiescent state for even the tiniest fraction of a second, the pull of gravity of one region on another would have been strong enough to force everything together into what is called a black hole. We know that didn't happen; otherwise we wouldn't be here to discuss the event.

Let me explain. Gravity moves with the speed of light. That means that if the sun somehow got heavier just now, we wouldn't know about it for a few minutes (the time that light and gravity take to traverse the distance between the sun and the earth). In the same way, when the universe was just one billionth of a second old, each object could only exert a pull upon material within one foot of its location (and feel its pull in return); light travels just one foot in a billionth of a second. At one second after the birth of the universe, each object was being tugged at by the pull of material within a 186,000-mile radius, but by that time their tremendous initial speeds had already moved the pieces far enough apart to lower the density safely below the critical amount needed to form a black hole.

In order to achieve consistency with our observations we must, according to Einstein's General Relativity, assume not only creation of matter and energy out of nothing, but creation of space and time as well. Moreover, this creation must be very delicately balanced. The amount of energy given to the emerging matter must be enough to move it fast enough to escape the bonds of gravity, but not so fast that the particles lose all contact with each other. Enough of the initially-created matter must pull together under gravity to form galaxies, stars, and planetary systems which allow for life. Thus the second "improbable" property of the early universe, almost as improbable as creation out of nothing, is an exquisitely delicate balance between matter and energy. Third—and this one puzzles scientists at least as much as the first two—somehow all these pieces, each without having any proper contact with the others, without having any way of communication, *all* must have appeared with the same balance between matter and energy at the same instant.

Today, whenever we stand bareheaded under the open sky, our scalps absorb a tiny portion of the heat left over from the Big Bang. This is the "background radiation" that R.W. Wilson and I discovered some years ago, radio waves which are now arriving at the earth from a distance of some eighteen billion light years. These radio photons also add a little bit to the "snow" on your television set, as well as to the whooshing sound you hear between stations on your FM receiver. These photons started their flight toward us almost eighteen billion years ago. Until they arrived here we had no communication with the part of the universe they came from. (Yesterday's photons came from a slightly closer region.) As time goes on, more of the pieces of the universe gradually come into contact with one another. Now it turns out that all regions of the universe we have seen so far appear identical in form, and in their laws of nature, to the other parts—even though they hadn't been in contact when their composition was established.

Some of you may have read about proposed modifications to the Big Bang theory such as the "bubble theory". Most have to do with hypotheses for how this universal perfection could have happened without violating our understanding of the laws of physics. The bubble theory is a mathematical attempt at getting around our third "improbable" observational fact. As of now, the attempt seems to have been unsuccessful, but the importance of the challenge suggests that scientists will continue to pursue such lines of attack.

Before concluding, I can't resist bringing up the "missing mass," the difference between the amount of matter astronomers find in the universe and the much larger amount needed to reverse the flight of the galaxies (and ultimately pull them back into a single condensed state). Naively, one might imagine hunting for matter as a kind of astronomical inventory, one in which the total climbs as overlooked nooks and crannies are examined. No way! Just as astronomers "weigh" the sun by measuring the motion of the earth, we infer the mass of the universe from the motion of the galaxies themselves. Those motions point to a universe which will fly apart indefinitely—*not* one which will someday collapse to a point. Thus observations also contradict the notion that our Big Bang is just one of an infinite series of such events.

Astronomy leads us to a unique event, a universe which was created out of nothing, one with the very delicate balance needed to provide exactly the conditions required to permit life, and one which has an underlying (one might say "supernatural") plan. Thus, the observations of modern science seem to lead to the same conclusions as centuries-old intuition. At the same time, most of our modern scientific intuition seems to be more comfortable with the world as described by the science of yesterday. Kind of interesting, isn't it?

· · · · · · · · · ·

Note

Quoted from *Guide of the Perplexed of Maimonides*, New York: Hebrew Publishing Co., 1946, Part II, Ch. XXX. I am deeply indebted to Jacob Dienstag for stimulating discussions and helpful material on this subject.

17

Science Asks What and How, While Religion Asks Why

• • • • • • • • • •

Professor John G. Phillips

• Born 9 January 1917

• Ph.D. in astronomy, University of Chicago, 1948

• Currently Professor of Astronomy, University of California, Berkeley

• Areas of specialization and accomplishments: laboratory studies of molecules of interest to astrophysics and the application of this information to study of spectra of cooler stars; works include (with S. P. Davis) *The Red System of the CN Molecule*, 1963

• Professor Phillips on:

the origin of the universe: ". . . there is solid evidence that the universe is some fifteen billion years old".

the origin of life and of *Homo sapiens*: A result of "appropriate conditions".

God: ". . . there seems to be some force influencing the evolution of societies; to deny its existence is to deny any purpose to life".

• • • • • • • • • •

1 What do you think should be the relationship between religion and science?

I see no conflict between science and religion; they are basically addressing different questions. Science asks what and how, while religion asks why.

2 What is your view on the origin of the universe: both on a scientific level and—if you see the need—on a metaphysical level?

While details are still to be worked out, there is solid evidence that the universe is some fifteen billions years old.

3 What is your view on the origin of life: both on a scientific level and—if you see the need—on a metaphysical level?

There is no reason to believe that, given the appropriate conditions (time, temperature, ingredients, and so forth), life would not evolve anywhere in the universe.

4 What is your view on the origin of *Homo sapiens*?

As far as human life is concerned, my answer would be the same as for Question 3.

5 How should science—and the scientist—approach origin questions, specifically the origin of the universe and the origin of life?

See my answers to Questions 2 and 3 above.

6 Many prominent scientists—including Darwin, Einstein, and Planck—have considered the concept of God very seriously. What are your thoughts on the concept of God and on the existence of God?

I have difficulty visualizing a disembodied individual "God" sitting in some hypothetical "heaven" governing human affairs. Yet there seems to be some force influencing the evolution of societies; to deny its existence is to deny any purpose to life.

18
Temporal Origin and Ontological Origin
• • • • • • • • • •
Professor John Polkinghorne

• Born 16 October 1930

• Ph.D. in physics, Cambridge University, 1955

• Currently Dean, Trinity Hall, and President, Queens' College, Cambridge University

• Areas of specialization and accomplishments: analytic side of elementary particle physics, including the analytic and high-energy properties of Feynman integrals and the foundations of S-Matrix theory; works include numerous papers on theoretical elementary particle physics in scientific journals; *The Analytic S-Matrix*, 1966; *The Particle Play*, 1979; *Models of High Energy Processes*, 1980; *The Quantum World*, 1984

• Professor Polkinghorne on:

the origin of the universe: "Theology is not concerned with temporal origin but ontological origin; creation is not an act of the remote past but a continuing act of the divine will in every present moment".

the origin of life: A result of "the astonishing potentiality with which matter-in-flexible-organization is endowed".

the origin of *Homo sapiens:* "I see humankind as qualitatively different from the animals because of its self-consciousness and its ability to know and worship its Creator".

God: "I believe that God exists . . ."

• • • • • • • • • •

These are matters about which one tries to write books (*One World; Science and Creation; Science and Providence*) but here is a brief response:

1 What do you think should be the relationship between religion and science?

> I think that science is autonomous within its self-limited domain and that theology (the intellectual reflection upon religion) is the great integrating discipline that sets the results of other human enquiry within the most profound and comprehensive matrix of understanding.

2 What is your view on the origin of the universe: both on a scientific level and—if you see the need—on a metaphysical level?

> I accept the story of the cosmologists, tracing back the history of our universe to within a fraction of a second of the apparent "Big Bang", with the necessary reserve about the more speculative assertions of what happened at the very earliest epochs. It is even conceivable that the whole show originated from the vacuum. However, this would not be creation *ex nihilo*, nor would it answer Leibniz's great metaphysical question, "Why is there something rather than nothing?" I accept the theistic doctrine of God the Creator, the One who holds the world in being. Theology is not concerned with temporal origin but ontological origin; creation is not an act of the remote past but a continuing act of the divine will in every present moment.

3 What is your view on the origin of life: both on a scientific level and—if you see the need—on a metaphysical level?

> I believe that living creatures have evolved from inanimate matter through a continuous process of development, at present not well understood scientifically. Metaphysically this speaks to me of the astonishing potentiality with which matter-in-flexible-organization is endowed. Indeed, I believe that our mental experience is a complementary pole to our material experience. In other words I seek some non-reductionist form of monism (neither materialism nor idealism).

4 What is your view on the origin of *Homo sapiens*?

> I think that humanity has arisen from lower forms of life without the injection of an extra "spiritual" ingredient. Nevertheless, because I believe that evolving complexity brings about genuine novelty, I see humankind as qualitatively different from the animals because of its self-consciousness and its ability to know

and worship its Creator. In that sense we are indeed spiritual beings, but inescapably embodied (not apprentice angels).

5 How should science—and the scientist—approach origin questions, specifically the origin of the universe and the origin of life?

I believe that, in principle, scientifically posable questions are scientifically answerable. We should use our scientific knowledge and abilities to learn all we can about the probable early history of the universe and about how inanimate matter complexified into living matter. However, other questions which we surely must ask—such as, Why is there a world at all? Why is it the way it is in its given law and circumstance? Is there a purpose behind cosmic history?—are not scientific and require metaphysics for their answer. I find the most satisfying and comprehensive answers to be provided by theism.

6 Many prominent scientists—including Darwin, Einstein, and Planck—have considered the concept of God very seriously. What are your thoughts on the concept of God and on the existence of God?

I take God very seriously indeed. I am a Christian believer (indeed, an ordained Anglican priest), and I believe that God exists and has made himself known in human terms in Jesus Christ.

19

I Have Difficulty Accepting that Matter Has Been in Existence Forever

· · · · · · · · · ·

Professor John A. Russell

- Born 23 March 1913

- Ph.D. in astronomy, University of California, 1943

- Distinguished Emeritus Professor of Astronomy, University of Southern California

- Areas of specialization: meteor spectroscopy and statistics

- Professor Russell on:

  the origin of the universe: "Frankly, I have difficulty conceiving spontaneous creation in the distant past as a scientific process or accepting that matter has been in existence forever".

  the origin of life: Darwin's theories seem to be "supported by an impressive amount of observational evidence".

  the origin of *Homo sapiens:* ". . . man has evolved as a consequence of a fork developing in the evolutionary track".

  God: "I believe that the universe was created and is sustained by some power that we call God".

· · · · · · · · · ·

1 What do you think should be the relationship between religion and science?

There once was a poster on the wall near my office that showed two urchins painting a very ramshackle playhouse. Random swaths of blue paint crisscrossed what passed as walls. The title under the poster read, "The building reveals the builder". I believe science is the source of much information about the universe and our place in it that reveals its creator. I was raised in a very liberal Congregational church where I was encouraged to explore and reconcile my science and my religion. I feel sorry for those who seek security in religious dogma that becomes more and more at odds with science. Science should not replace religion, but religion that flagrantly violates science is generally untenable for me. I acknowledge freely that religion goes beyond science in that it grapples with timeless questions that science cannot answer. Russell (alas! no relative of mine), Dugan, and Stuart close the introduction of their classical, two-volume text with these cogent lines:

> Since astronomy is a physical science, it can give no direct answer to problems in philosophy. Thus, while it can tell that a star is larger than a man, it cannot decide which possesses the greater worth.
>
> Nevertheless, the realization which this science brings of the tremendous extent of the material universe in space and time, and of its essential unity, in that the same types of matter and the same natural laws are found everywhere, is of great significance. Though our planet thus appears as an insignificant speck, it is yet likely to be habitable for millions of years to come. The appreciation of these facts cannot fail to possess an important influence in determining the attitude of the contemplative student towards such problems of philosophy as man's obligations to future generations, his place in the universe, and his relation to the Power which is behind it.

2 What is your view on the origin of the universe: both on a scientific level and—if you see the need—on a metaphysical level?

Frankly, I have difficulty conceiving spontaneous creation in the distant past as a scientific process or accepting that matter has been in existence forever. As the universe shows convincing evidence of evolution in the last fifteen billion years or so, I should lean toward spontaneous creation of some sort at some time. As to what preceded the Big Bang I could only speculate. I wonder if the theoretical astrophysicists may not have carried their mathematical extrapolations beyond a point verifiable by any physical reality. Between a time of extraordinary concentration of matter billions of years ago and today, physics and some form of the Kant-Laplace concept answer most of the questions.

3 What is your view on the origin of life: both on a scientific level and—if you see the need—on a metaphysical level?

From my first exposure to them, Darwin's theories seemed to me to be supported by an impressive amount of observational evidence. Reason supplies no support in my judgment for the fundamentalist objection to evolution as denigrating the power of God. If I were God and had thought of evolution as a process for developing life suitable to its environment, I should have considered evolution as a remarkable way to do it.

4 What is your view on the origin of *Homo sapiens*?

I accept the concept that man has evolved as a consequence of a fork developing in the evolutionary track, where one branch of the fork led to land dwellers (humans) and the other branch led to tree dwellers (apes).

5 How should science—and the scientist—approach origin questions, specifically the origin of the universe and the origin of life?

Although we have made great strides with science in pushing back the boundaries of the unknown, I am not convinced that science will take us to the threshold of complete understanding. Questions of origin and evolution imply that the universe is changing. The question is, is it upward change? I can't say that the universe is becoming better, except to the extent that it has resulted in the origin of at least one body on which life capable of intelligently studying time and space has developed. Is man really better than his predecessors? I may be falling into the trap of the seventeenth-century scientists who believed that all planets were inhabited because they could think of no other cause for their being created.

Today we ask, whither man? It is disquieting that we have found no shred of evidence that our counterparts exist elsewhere in the universe. Is it inevitable that the march of progress will be stopped by over-population or the accidental or deliberate misuse of atomic energy, both of which man has the power to control? Our ancestors survived many forces of nature beyond their power to control: fires, floods, earthquakes, plagues, and so on. We can only hope that if our civilization loses the battle for survival, others on other planets will make wiser decisions.

6 Many prominent scientists—including Darwin, Einstein, and Planck—have considered the concept of God very seriously. What are your thoughts on the concept of God and on the existence of God?

I believe that the universe was created and is sustained by some power that we call God. Through the ages man has pictured his gods as supermen. We assume that God thinks as we do and evaluates as we do. Great religions tend to center about a man whose sensitivity to the relation of man to God significantly exceeds that of lesser mortals and who may act as a go-between.

For me, the true nature of God is beyond my comprehension. I find some solace in the thought that he understands me better than I understand him. In his *Astronomy and the Cosmos*, Sir James Jeans poses questions including one that makes a huge leap beyond the boundaries of our customary inquiries. I close this comment on Question 6, as I customarily closed my elementary classes, with this direct quotation:

> The (astronomer) has finished his task when he has described to the best of his ability the inevitable sequence of changes which constitute the history of the material universe. But the picture which he draws opens questions of the widest interest not only to science, but also to humanity. What is the meaning, if any there be which is intelligible to us, of vast accumulations of matter which appear, on our present interpretations of space and time, to have been created only in order that they may destroy themselves? What is the relation of life to that universe, of which, if we are right, it can occupy so small a corner? What, if any, is our relation to the remote (galaxies), for surely there must be some more direct contact than that light can travel between them and us in a hundred million years? Do their colossal uncomprehending masses come nearer to representing the main ultimate reality of the universe, or do we? Are we merely part of the same picture as they, or is it possible that we are part of the artist? Are they perchance only a dream, while we are the brain-cells in the mind of the dreamer? Or is our importance measured solely by the fractions of space and time we occupy— space infinitely less than a speck of dust in a vast city, and time less than one tick of a clock which has endured for ages and will tick on for ages yet to come?

20

Science and Religion: Reflections on Transcendence and Secularization

••••••••••

Professor Abdus Salam

• Born 29 January 1926

• Ph.D. in theoretical physics, Cambridge University, 1952

• Nobel Prize for Physics (shared with Sheldon L. Glashow and Steven Weinberg), 1979; received the Nobel Prize with Glashow and Weinberg "for their contributions to the theory of the unified weak and electromagnetic interaction between elementary particles, including, inter alia, the prediction of the weak neutral current"; works include *Symmetry Concepts in Modern Physics*, 1966; *Ideals and Realities: Selected Essays*, 1984

• Currently Director, International Centre for Theoretical Physics, Trieste (Italy), and President, Third World Academy of Sciences

• Professor Salam:

"Now this sense of wonder leads most scientists to a Superior Being— der Alte, the Old One, as Einstein affectionately called the Deity—a Superior Intelligence, the Lord of all Creation and Natural Law".

••••••••••

Science as Anti-Religion

It is generally stated that science is anti-religion and that science and religion battle against each other for the minds of men. Is this correct?[1]

Now if there is one hallmark of true science, if there is one perception that scientific knowledge heightens, it is the spirit of wonder; the deeper that one goes, the more profound one's insight, the more is one's sense of wonder increased. This sentiment was expressed in eloquent verse by Faiz Ahmad Faiz:

Moved by the mystery it evokes, many a time have I dissected the heart of the smallest particle. But this eye of wonder; its wonder-sense is never assuaged!

93

In this context, Einstein, the most famous scientist of our century, has written:

> The most beautiful experience we can have is of the mysterious. It is the fundamental emotion which stands at the cradle of . . . true science. Whoever does not know it and can no longer wonder, no longer marvel, is as good as dead, and his eyes are dimmed. It was the experience of mystery—even if mixed with fear—that engendered religion. A knowledge of the existence of something we cannot penetrate, our perceptions of the profoundest reason and the most radiant beauty, which only in their most primitive forms are accessible to our minds—it is this knowledge and this emotion that constitute true religiosity; in this sense, and in this alone, I am a deeply religious man.

Einstein was born into an Abrahamic faith; in his own view, he was deeply religious.

Now this sense of wonder leads most scientists to a Superior Being—der Alte, the Old One, as Einstein affectionately called the Deity—a Superior Intelligence, the Lord of all Creation and Natural Law. But then the differences start, and let us discuss these.

The Abrahamic religions claim to provide a meaning to the mystery of life and death. These religions speak of a Lord who not only created natural law and the universe in his glory, his own holiness and his majesty; but also created *us*, the human beings in his own image, endowing us not only with speech, but also with spiritual life and spiritual longings. This is one aspect of transcendence. The second aspect is of the Lord who answers prayers when one turns to him in distress. The third is of the Lord who, in the eyes of the mystic and the Sufi, personifies eternal beauty and is to be adored for this. These transcendent aspects of religion as a rule lead to a heightening of one's obligation towards living beings. The fourth is of the Lord who endows some humans—the prophets and his chosen saints—with divinely inspired knowledge through revelation.

Regarding what may be called (in the present context) the "societal", "secularist" thinking, Abrahamic religions speak of the Lord who is also the guardian of the moral law, the precept which states that "Like one does, one shall be done by"; the Lord who gives a meaning to the history of mankind—the rise and fall of nations for disobedience to his commandments; the Lord who specifies what should be human belief as well as ideal human conduct of affairs[2]; and finally, the Lord who rewards one's good deeds and punishes wrongdoing (like a father), in this world or a life hereafter.

While many scientists in varying degrees do subscribe to the first three aspects of transcendentalism, not many subscribe to the "societal" aspects of religiosity.[3] Scientists have their own dilemmas in this respect.

. . . .

The Three Viewpoints of Science

Let us start with natural law, which governs the universe. There are scientists who would take issue with Einstein's view that there is a sublime beauty about the laws of nature and that the deepest (religious) feelings of man spring from the sense of wonder evoked by this beauty. These scientists would instead like to deduce the laws of nature from a self-consistency and "naturalness" principle, which made the universe come into being spontaneously. This should be something like the doctrine of spontaneous creation of life and its Darwinian evolution, only now carried to the realm of all laws of nature and the whole universe. If successful, this, in their view, would lead to an irrelevance of a deity.[4] Man's spiritual dimension, so called, would be nothing but a particular manifestation of physiological processes occurring inside the human brain (not fully understood at present), but their hope would be that a molecular basis would one day be discovered for this.[5]

Contrasting with this is the view of the anthropic scientist who believes that the universe was created purposefully with such attributes and in such a manner that sentient beings could arise. These then are the three viewpoints—first, the (religious and transcendental) attitude of an Einstein; second, the anthropic view (which in a way supports the first); and third, the viewpoint of the self-consistent scientist in whose scheme of things the concept of a Lord is simply irrelevant.

Regarding what I have called the secularist sentiments in general, Einstein has this to say: "I am satisfied with the mystery of the eternity of life and with the awareness and a glimpse of the marvelous structure of the existing world, together with the devoted striving to comprehend a portion, be it ever so tiny, of the Reason that manifests itself in nature. . . . (But) I cannot conceive of a God who rewards and punishes his creatures, or has a will of the kind that we experience in ourselves. . . . The existence and validity of human rights are not written in the stars". Instead his belief was that "the ideals concerning the conduct of men toward each other and the desirable structure of the community have been conceived and taught by enlightened individuals in the course of history".

Apart from the subjective character of the opinion, note Einstein's silence about the spiritual dimension of religion.

• • • • • • • • • •

Modern Science and Faith

In the formation of such attitudes toward religion, could it be that the medieval Church was partly responsible through its opposition to science? Could it be that these attitudes are a legacy of the battles of yesterday when the so-called "rational philosophers", with their irrational and dogmatic faith in the cosmological doctrines they had inherited from Aristotle, found difficulties in reconciling these with their faith?

One must remind oneself that the battle of faith and science was fiercely waged among the schoolmen of the Middle Ages. The problems which concerned the schoolmen were mainly problems of cosmology: "Is the world located in an immobile place? Does anything lie beyond it? Does God move the *primum mobile* directly and actively as an efficient cause or only as a final or ultimate cause? Are all the heavens moved by one mover or several? Do celestial movers experience exhaustion or fatigue? What is the nature of celestial matter? Is it like terrestrial matter in possessing inherent qualities such as being hot, cold, moist, and dry?" When Galileo tried, first, to classify those among the problems which legitimately belonged to the domain of physics, and then to find answers, only to them, through physical experimentation, he was persecuted. Restitution for this is, however, being made now, three hundred and fifty years later.

At a special ceremony in the Vatican on 9 May 1983, His Holiness the Pope declared: "The Church's experience, during the Galileo affair and after it, has led to a more mature attitude. . . . The Church herself learns by experience and reflection and she now understands better the meaning that must be given to freedom of research . . . one of the most noble attributes of man. . . . It is through research that man attains to Truth. . . . This is why the Church is convinced that there can be no real contradiction between science and faith. . . . (However,) it is only through humble and assiduous study that (the Church) learns to dissociate the essential of the faith from the scientific systems of a given age, especially when a culturally influenced reading of the Bible seemed to be linked to an obligatory cosmogony".

· · · · · · · · · ·

The Limitations of Science

In the remarks I have quoted, His Holiness the Pope stressed the maturity which the Church had reached in dealing with science; he could equally have emphasized the converse—the recognition by scientists, from Galileo's times onwards, of the limitations of their disciplines—the recognition that there are questions which are beyond the ken of present (or even future) science and that "science has achieved its success by restricting itself to a certain type of inquiry".

We may speculate about some of them, but there may be no way to verify empirically our metaphysical speculations. And it is this empirical verification that is the essence of modern science. We are humbler today than, for example, Ibn Rushd (Averroes) was. Ibn Rushd was a physician of great originality with major contributions to the study of fevers and of the retina; this is one of his claims to scientific immortality. However, in a different scientific discipline—cosmology—he accepted the speculations of Aristotle, without recognizing that these were speculations and that the future might prove Aristotle wrong. The scientist of today knows when and where he is speculating; he would claim no finality even for the associated modes of thought. And about accepted facts, we recognize that newer facts may be discovered which, without falsifying the earlier discoveries, may lead to generalizations; in turn, necessitating revolutionary changes in our "concepts" and our "worldview". In physics, this happened twice in the beginning of this century, first with the discovery of relativity of time and space, and second with quantum theory. It could happen again, with our present constructs appearing as limiting cases of the newer concepts—still more comprehensive, still more embracing.

Permit me to elaborate on this.

I have mentioned the revolution in the physicists' concepts of the relativity of time. It appears incredible that the length of a time interval depends on one's speed—that the faster we move the longer we appear to live to someone who is not moving with us. And this is not a figment of one's fancy. Come to the particle physics laboratories of CERN at Geneva which produce short-lived particles like muons, and make a record of the intervals of time which elapse before muons of different speeds decay into electrons and neutrinos. The faster muons take longer to die, the slower ones die and decay early, precisely in accord with the quantitative law of relativity of time first enunciated by Einstein in 1905.

Einstein's ideas on time and space brought about a revolution in the physicist's thinking. We had to abandon our earlier modes of thought in physics. In this context, it always surprises me that the professional philosopher and the mystic—who up to the nineteenth century used to consider space and time as their special preserve—have somehow failed to erect any philosophical or mystical systems based on Einstein's notions!

The second and potentially the more explosive revolution in thought came in 1926 with Heisenberg's Uncertainty Principle. This principle concerns the existence of a conceptual limitation on our knowledge. It affirms, for example, that no physical measurements can tell you simultaneously that there is an electron on the table—here—and also that it is lying still. Experiments can be made to discover precisely where the electron is; these same experiments will then destroy any possibility of finding at the same time whether the electron is moving and if so at what speed. There is an inherent limitation on our knowledge, which appears to have been "decreed in the nature of things".[6] I shudder to think what might have happened to Heisenberg if he was born in the Middle Ages—just what theological battles might have raged on whether there was a like limitation on the knowledge possessed by God.

As it was, battles were fought, but within the twentieth-century physics community. Heisenberg's revolutionary thinking, supported by all known experiments, has not been accepted by all physicists. The most illustrious physicist of all times, Einstein, spent the best part of his life trying to find flaws in Heisenberg's arguments. He could not gainsay the experimental evidence, but he hoped that such evidence might perhaps be explained within a different "classical" theoretical framework. Such framework has not been found so far. Will it ever be discovered?

. . . .

Faith and Science

But is the science of today really on a collision course with metaphysical thinking? The problem, if any, is not peculiar to the faith of Islam—the problem is one of science and faith in general, at least so far as the Abrahamic religions are concerned. Can science and faith at the least live together in "harmonious complementarity"? Let us consider some relevant examples of modern scientific thinking.

My first example concerns the "metaphysical" doctrine of creation from nothing. Today, a growing number of cosmologists believe that the most likely value for the density of matter and energy in the universe is such that the mass of the universe adds up to zero, precisely. The mass of the universe is defined as the sum of the masses of matter—the electrons, the protons, and the neutrinos, which constitute the universe as we believe we know it—plus their mutual gravitational energies (converted into mass; the gravitational energy of an attractive force is negative in sign). If the mass of the universe is indeed zero—and this is an empirically determinable quantity—then the universe shares with the vacuum state the property of masslessness. A bold extrapolation made around 1980 then treats the universe as a quantum fluctuation of the vacuum—of the state of nothingness.

Attractive idea, but at the present time, measurements do not appear to sustain it. This has led to an ongoing search for a new type of matter—the so-called "dark matter"—which is not luminous to us but would show itself to us only through its gravity.

We shall soon know empirically whether such matter exists or not. If it does not, we shall discard the whole notion of the universe arising as a quantum fluctuation. This may be a pity, but this points to a crucial difference between physics and metaphysics—experimental verification is the final arbiter of even the most seductive ideas in physics.

· · · · · · · · · ·
Anthropic Universe

My second example is the principle of the anthropic universe—the assertion by some cosmologists that one way to understand the processes of cosmology, geology, biochemistry, and biology is to assume that our universe was conceived in a potential condition and with physical laws which possess all the necessary ingredients for the emergence of life and intelligent beings. "Basically this potentiality relies on a complex relationship between the expansion and the cooling of the Universe after the Big Bang, on the behaviour of the free energy of matter, and on the intervention of chance at various (biological) levels", as well as on a number of "coincidences" which, for example, have permitted the universe to survive the necessary few billion years.

Stephen Hawking, the successor of Newton in the Lucasian Chair at Cambridge, in his recent book, *A Brief History of Time: From the Big Bang to Black Holes* (New York: Bantam Press, 1988), has stated the anthropic principle most succinctly:

> There are two versions of the anthropic principle, the weak and the strong. The weak anthropic principle states that in a universe that is large or infinite in space and/or time, the conditions necessary for the development of intelligent life will be met only in regions that are limited in space and time. The intelligent beings in these regions should therefore not be surprised if they observe that their locality in the universe satisfies the conditions that are necessary for their existence. It is a bit like a rich person living in a wealthy neighbourhood not seeing any poverty.
>
> One example of the use of the weak anthropic principle is to "explain" why the big bang occurred about ten thousand million years ago—it takes about that long for intelligent beings to evolve. . . . An early generation of stars first had to form. These stars converted some of the original hydrogen and helium into elements like carbon and oxygen, out of which we are made. The stars then exploded as supernovas, and their debris went to form other stars and planets, among them those of our solar system, which is about five thousand million years old. The first one or two thousand million years of the earth's existence were too hot for the development of anything complicated. The remaining three thousand million years or so have been taken up by the slow process of biological evolution, which has led from the simplest organisms to beings who are capable of measuring time back to the big bang.
>
> Few people would quarrel with the validity or utility of the weak anthropic principle. Some, however, go much further and propose a strong version of the principle. According to this theory, there are either many different universes or many different regions of a single universe, each with its own initial configuration and, perhaps, with its own set of laws of science. In most of these universes the conditions would not be right for the development of complicated organisms; only in the few universes that are like ours would intelligent beings develop and ask the question: "Why is the universe the way we see it?" The answer is then simple: if it had been different, we would not be here!
>
> The laws of science, as we know them at present, contain many fundamental numbers, like the size of the electric charge

of the electron and the ratio of the masses of the proton and the electron. We cannot, at the moment at least, predict the values of these numbers from theory—we have to find them by observation. It may be that one day we shall discover a complete unified theory that predicts them all, but it is also possible that some or all of them vary from universe to universe or within a single universe. The remarkable fact is that the values of these numbers seem to have been very finely adjusted to make possible the development of life. For example if the electric charge of the electron had been only slightly different, stars either would have been unable to burn hydrogen and helium, or else they would not have exploded. Of course, there might be other forms of intelligent life, not dreamed of even by writers of science fiction, that did not require the light of a star like the sun or the heavier chemical elements that are made in stars and are flung back into space when the stars explode. Nevertheless, it seems clear that there are relatively few ranges of values for the numbers that would allow the development of any form of intelligent life. Most sets of values would give rise to universes that, although they might be very beautiful, would contain no one able to wonder at that beauty. One can take this either as evidence of a divine purpose in Creation and the choice of the laws of science or as support for the strong anthropic principle.

Another example of the anthropic principle at work is provided by the recently-discovered "electroweak" force. It is interesting to ask why nature has decided to unify the electromagnetic and weak nuclear forces into one electroweak force. (The electroweak together with the strong nuclear and the gravitational forces constitute the three fundamental forces that we know about in nature.) One recent answer to this question seems to be that this unification provides one way to understand why in the biological regime one finds amino acids which are only left-handed and sugars which are only right-handed. (Left- and right-handedness refers to the polarization of light after its scattering from the relevant molecules.) In the laboratory, both types of molecules, left-handed as well as right-handed, are produced in equal numbers. Apparently over biological time, one type of handed molecule decayed into the other type.

According to some scientists, the handedness of naturally occurring molecules is predicated by the fact that electromagnetism (the force of chemistry) is unified with the weak nuclear force—a force which is well-known to be handed (for example, the weak nuclear force exists only between left-handed neutrons and left-handed electrons). This fact, plus the long biological time available for life to emerge, apparently was responsible for the observed handedness of biological molecules in neutrons.

But where does the anthropic principle come into this? One indication of this could be as follows: Take penicillin as an example. Bacteria, exceptionally, utilize right-handed D-amino acids in the construction of their cell walls. Penicillin itself contains a group of right-handed amino acids and interferes with the syntheses of the bacterial cell walls, a process which is unique to bacteria and which does not occur in the mammalian host. The penicillin miracle would thus be impossible except for the unification of electromagnetism and weak forces. (Why one would have to wait for ten billion years before sentient beings came into existence to marvel, is of course not explained by the anthropic principle.)

• • • • • • • • • •

The Self-Consistency Principle

Finally, there is the third category of scientists, those who use "self-consistency" and "naturalness" to explain the architecture of the universe. To illustrate self-consistency as applied in physics, I shall take a recent example.

As an extension of the recent excitement in physics—that is, of our success with the electroweak force, our success in unifying and establishing the identity of two of the fundamental forces of nature, the electric force and the weak nuclear forces—some of us are now seriously considering the possibility that space-time may have ten dimensions. Within this context we hope to unify the electroweak force with the remaining two basic forces, the force of gravity and the strong nuclear force. This is being done nowadays (1988) as part of "supersymmetric string theories" in ten dimensions. The attempt, if successful, will present us with a unified "Theory of Everything".

Of the ten dimensions, four are the familiar dimensions of space and time. The other six dimensions are supposed to correspond to a hidden internal manifold—hidden because these six dimensions are assumed to have curled in upon themselves to fantastically tiny dimensions of the order of 10^{-33} cms. We live in a six-dimensional manifold in the ten-dimensional space-time: Our major source of sensory apprehension of these extra dimensions is the existence of familiar electric charges—electric and nuclear, which in their turn produce the familiar electric and the nuclear forces.

Exciting idea, which may or may not work out quantitatively. But one question already arises: why the difference between the four familiar space-time dimensions and the six internal ones?

So far our major "success" has been in the understanding of why ten dimensions in the first place (and not a wholesome number of dimensions like thirteen or nineteen). This apparently has to do with the "quantum anomalies" which plague the theory (and produce unwanted infinities) in any but ten dimensions. The next question which will arise is this: Were all the ten dimensions on par with each other at the beginning of time? Why have the six curled in upon themselves, while the other four have not?

The unification implied by the existence of these extra dimensions curling in upon themselves is one of the mysteries of our subject. At present, we would like to make this plausible by postulating a "self-consistency and naturalness" principle. (This has not yet been accomplished.) But even if we are successful, there will be a price to pay; there will arise subtle physical consequences of such self-consistency—for example, possibly remnants, just like the three-degrees Kelvin radiation which we believe was a remnant of the recombination era following on the Big Bang. We shall search for such remnants. If we do not find them, we shall abandon the idea.

Creation from nothing, extra and hidden dimensions—strange topics for late twentieth-century physics, which appear no different from the metaphysical preoccupations of earlier times; however, they are all driven by a self-consistency principle. So far as physics is concerned, mark however the insistence on empirical verification at each stage.

• • • • • • • • • •

Notes

1. One must recognize at the outset that religion is one of the strongest "urges of mankind", which can make men and women sacrifice their all, including their lives, for its sake.

2. A Jew like Einstein was Jewish because he subscribed to the ostensibly "cultural aspects" of the Jewish faith, rather than any "fundamentalist" belief in the teachings regarding "ideal human conduct" in the Old Testament. Freud expressed himself, in a similar vein, in his preface to the Hebrew translation of *Totem and Taboo*. Referring to the

emotional position of an author (himself) who is ignorant of the language of holy writ and estranged from the religion of his fathers, he says that if the question were put to him, "Since you have abandoned all these common characteristics . . . what is there left to you that is Jewish?" he would reply, "A very great deal, and probably its very essence." He said he "could not now express that essence clearly in words which someday, no doubt, would become accessible to the scientific mind".

3. The specification of "ideal belief and conduct" unfortunately has almost always led to intolerance, excommunication, fanaticism, and repression, particularly of minorities. *There is divisiveness in the very concept of the chosen people.* In its worst manifestations this divisiveness may sanction murder for religious disagreements, often making a mockery of a religion's own tolerant teachings. In this respect, the late Professor Sir Peter Medawar, Nobel Laureate in Biology and Medicine, in his book *The Limits of Science* (Oxford University Press, 1980) had this to say: "Religious belief gives a spurious spiritual dimension to tribal enmities. . . . The only certain way to cause a religious belief to be held by everyone is to liquidate nonbelievers. . . . The price in blood and tears that mankind generally has had to pay for the comfort and spiritual refreshment that religion has brought to a few has been too great to justify our entrusting moral accountancy to religious belief. By 'moral accountancy' I mean the judgment that such-and-such an action is right or wrong, or such a man good and such another evil."

4. I find the creationist creed especially insulting in that while we ascribe subtlety to ourselves in devising these self-consistency modalities, the only subtlety we are willing to ascribe to the Lord is that of a potter's art—kneading clay and fashioning it into man.

5. Since the twentieth century has been called "the century of science", I wish I could somehow convey the depth of the miracle of modern science, both basic and applied. The twentieth century has been a century of great syntheses in science—the syntheses represented by quantum theory, relativity, and unification theories in physics, by the Big Bang idea in cosmology, by the genetic code in biology, and by ideas of plate tectonics in geology; likewise in technology, the conquest of space and the harnessing of atomic power. Just as in the sixteenth century when European man discovered new continents and occupied them, the frontiers of science are being conquered one after another. I have always felt passionately that our men and women in Arab-Islamic lands should also be in the vanguard of making these conquests today, as they were before the year 1500.

6. One of the most difficult questions which the self-consistent scientist has to answer is: Why this decree?

21

One Must Ask Why and Not Just How

Professor Arthur L. Schawlow

- Born 5 May 1921

- Ph.D. in physics, University of Toronto, 1949; Nobel Prize for Physics (shared with Nicolaas Bloembergen and Kai Siegbahn), 1981; received the Nobel Prize with Bloembergen "for their contribution to the development of laser spectroscopy"

- Currently J.G. Jackson-C.J. Wood Professor of Physics at Stanford University

- Professor Schawlow on:

  the origin of the universe: ". . . the ultimate origin of the universe may be not only unknown but unknowable. . . . there is no real way to find out what came before the Big Bang".

  the origin of life: Even if the origin of life can eventually be broken down "to a series of chemical steps, subject to known physical laws, it will still be marvelous that those powerful laws have such enormous potential".

  the origin of *Homo sapiens*: ". . . there is excellent evidence that evolution did occur . . . But whatever is found can be understood as God's way of producing humans".

  God: "It seems to me that when confronted with the marvels of life and the universe, one must ask why and not just how. The only possible answers are religious. . . . I find a need for God in the universe and in my own life".

1 What do you think should be the relationship between religion and science?

Science cannot either prove or disprove religion. Religion is founded on faith. It seems to me that when confronted with the marvels of life and the universe, one must ask why and not just how. The only possible answers are religious. For me that means Protestant Christianity, to which I was introduced as a child and which has withstood the tests of a lifetime.

But the context of religion is a great background for doing science. In the words of Psalm 19, "The heavens declare the glory of God and the firmament showeth his handiwork". Thus scientific research is a worshipful act, in that it reveals more of the wonders of God's creation.

2 What is your view on the origin of the universe: both on a scientific level and—if you see the need—on a metaphysical level?

Current research in astrophysics seems to indicate that the ultimate origin of the universe may be not only unknown but unknowable. That is, if we assume the Big Bang which present evidence strongly supports, there is no real way to find out what came before the Big Bang. It is surely right to pursue as far as possible the scientific understanding of the origins of the universe, but it is probably wrong to think that we have final answers and that there are no further surprises to come. From a religious point of view, we assume that God did it and hope to find out something about how he did it.

3 What is your view on the origin of life: both on a scientific level and—if you see the need—on a metaphysical level?

The origin of life is also a fit subject for scientific enquiry. Even if it turns out that we can eventually break it down to a series of chemical steps, subject to known physical laws, it will still be marvelous that those powerful laws have such enormous potential.

4 What is your view on the origin of *Homo sapiens*?

My answer to this question is essentially the same as to the preceding one. Although my biological knowledge is not very deep, it seems that there is excellent evidence that evolution did occur and is still occuring. Scientists in that field should discover all they can about the process. But whatever is found can be understood as God's way of producing humans.

5 How should science—and the scientist—approach origin questions, specifically the origin of the universe and the origin of life?

Origin questions should be pursued as vigorously as the scientists' abilities and interests can take them. But the answers will never be final, and deeper questions will eventually have to be referred to religion.

6 Many prominent scientists—including Darwin, Einstein, and Planck—have considered the concept of God very seriously. What are your thoughts on the concept of God and on the existence of God?

As pointed out above, I find a need for God in the universe and in my own life. Some of the concepts of modern science provide useful metaphors for thinking about God. For instance, some scientists think of God as a sort of guiding principle, remote from the concerns of individuals. Yet they can freely use a time sharing computer, or the telephone switching network, which give essentially simultaneous attention to many individuals.

The ideas of complementarity also help. It is not surprising that different religious individuals have such different views of God, for they have varying backgrounds and knowledge. We know in science that we can only describe or explain things in terms of other things. We also know that some things cannot be described completely, exhibiting different aspects under differing conditions. Thus it is not surprising that philosophers and peasants have different concepts of God. We are fortunate to have the Bible, and especially the New Testament, which tells us so much about God in widely accessible human terms, even though it also leaves us some things that are hard to understand.

22

The Origin of the Universe Does Not Seem to Me to Be a Scientific Question

• • • • • • • • • •

Professor Emilio Segre

- Born 1 February 1905

- Ph.D. in physics, University of Rome, 1928; Nobel Prize for Physics (shared with Owen Chamberlain), 1959; received the Nobel Prize with Chamberlain "for their discovery of the antiproton"

- Professor of Physics at the University of California, Berkeley, until his retirement in 1972; also Professor Emeritus at the University of Rome

- Works include *Nuclei and Particles: An Introduction to Nuclear and Subnuclear Physics,*1964; *From X-Rays to Quarks: Modern Physicists and Their Discoveries,* 1980

- Professor Segre on:

  the origin of the universe: "The origin of the universe, at present, does not seem to me to be a scientific question. Scientific theories are usually validated by experiment, consistency tests, and predictive power, all of which are hardly applicable to the origin of the universe".

  the origin of life: "As of now I do not see the need to go beyond physics and chemistry".

  the origin of *Homo sapiens:* ". . . *Homo sapiens* evolved from other primates, but I doubt that we know the details of this evolution and what drove it".

  God: Agrees with a statement (quoted here) from Einstein.

• • • • • • • • • •

1 What do you think should be the relationship between religion and science?

> I would keep the two separate. There are examples of great scientists that go from very religious (for instance, Faraday or Cauchy) to agnostics, to atheists. This shows that scientific ability is uncorrelated with religious opinion.

2 What is your view on the origin of the universe: both on a scientific level and—if you see the need—on a metaphysical level?

> I have no opinions on the *origins* of the universe. I have some knowledge of present cosmology, Big Bang, and so forth. I know that such theories are subject to change with time, although each one leaves a residue which is incorporated in the next one. The origin of the universe, at present, does not seem to me to be a scientific question. Scientific theories are usually validated by experiment, consistency tests, and predictive power, all of which are hardly applicable to the origin of the universe. On a metaphysical level each individual may have his own opinions; I do not know how to prove or disprove them.

3 What is your view on the origin of life: both on a scientific level and—if you see the need—on a metaphysical level?

> I suspect that one day or another it will be possible to synthesize a "living" thing, although it may take a long time. As of now I do not see the need to go beyond physics and chemistry. I feel the burden is on the other side.

4 What is your view on the origin of *Homo sapiens*?

> I believe *Homo sapiens* evolved from other primates, but I doubt that we know the details of this evolution and what drove it. Natural selection is a component of evolution, but it may not be the only one.

5 How should science—and the scientist—approach origin questions, specifically the origin of the universe and the origin of life?

> Scientists and science should consider the origin questions as subject to the same methodology they use for other problems. They should recognize, however, that they may not reach a "final" answer, whatever this means. Clearly they know a lot more now than they did a few hundred years ago. Keep the good work going. It may be objected that what we know is infinitesimal compared to what we do not know. This may be true, but I would not be bothered too much by it.

6 Many prominent scientists—including Darwin, Einstein, and Planck—have considered the concept of God very seriously. What are your thoughts on the concept of God and on the existence of God?

This is a difficult and long question and it starts with a definition of God. I have read several of Einstein's writings on the subject and I agree, roughly speaking, with what he says.

The first problem is that the word "God" has many meanings, one for each religion at least. The concept is fraught with ambiguities and anthropomorphic ideas, with moral implications, with the concepts of "soul", immortality, and so forth.

Einstein has written (Helen Dukas and Banesh Hoffmann, *Albert Einstein, The Human Side*, Princeton University Press, 1979, p. 142):

> I cannot imagine a personal God who influences immediately the actions of individual creatures or makes direct judgement of them. I cannot do it despite the fact the mechanistic causality of modern science is to a certain extent under doubt. My religious attitude involves a modest admiration of the infinitely exceeding spirit which reveals itself in the limited facts which we, with our limited and uncertain spirit, are able to recognize concerning reality. Morality is a most important feature for us, not for God. (Editor's translation—H. M.)

I agree with the entire statement. In addition religion has entirely different connotations for reason and for the emotions. The second aspect is quite important but irrational, as love.

23

The Universe Is Ultimately to Be Explained in Terms of a Metacosmic Reality

• • • • • • • • • •

Professor Wolfgang Smith

- Born 18 February 1930

- Ph.D. in mathematics, Columbia University, 1957

- Currently Professor of Mathematics, Oregon State University; faculty positions at Massachusetts Institute of Technology and the University of California, Los Angeles

- Professor Smith's research on the aerodynamics of diffusion fields provided the theoretical key to the solution of the re-entry problem for space flight.

- Professor Smith on:

  the origin of the universe: "I accept the so-called Big Bang hypothesis. . ."

  the origin of life: ". . . life and the natural species originate, symbolically speaking, 'at the Center' and evolve towards the periphery".

  the origin of *Homo sapiens:* ". . . man is special, not in the mode of his origination, but quite simply in what he is, in the archetype, namely, which he manifests".

  God: ". . . nothing is more evident, more certain, than the existence or reality of God".

• • • • • • • • • •

1 What do you think should be the relationship between religion and science?

"Science", Einstein once said, "deals with what *is*; religion with what ought to be"—but this is not at all our view. Religion, of course, exists and functions on many levels, as it ought to and must; and in its more familiar manifestations, at any rate, it is in fact largely concerned in one way or another with "what ought to be". But if we consider the religious phenomenon in its highest forms—as indeed we should if we would understand its essence—the picture changes. For then we find that religion deals not just with ethical norms and human consolations, but with reality, precisely, and that on a level which is normally inaccessible, to say the least. This, in any case, is the perennial claim; and I, for my part, can see no sound reason to doubt its validity.

It would seem, therefore, that Einstein's dictum needs ot be revised: it may indeed be religion, taken at its summit, that actually "deals with what *is*", in contrast to science, which by its nature is constrained to deal with "what appears to be" (under conditions stipulated by its own *modus operandi*).

Strictly speaking, there can be no "dialogue" between science and religion. It is doubtful that the truths of religion can be adequately explained on the level of scientific discourse, any more than a three-dimensional body can be made to fit into a plane; and the attempt is prone to "flatten" and thus destroy the very thing one pretends to render intelligible. This is typically what takes place, one fears, when so-called religious authorities begin to dialogue. Nothing is in fact more fatal to religion than the pretension to "demythologize" its content.

What the scientist (like everyone else) needs in the face of the religious phenomenon is a profound humility. To understand what religion is one must first of all be religious oneself; the essential thing simply cannot be known from the outside.

2 What is your view on the origin of the universe: both on a scientific level and—if you see the need—on a metaphysical level?

On a scientific level I accept the so-called Big Bang hypothesis as a cogent and reasonably well substantiated theory. What renders that "model" all the more plausible, in my view, is the fact that it is clearly concordant with the traditional metaphysical cosmogonies. This is not the place to discuss the difference between the Judeo-Christian conception of *creatio ex nihilo* and the Platonic-Oriental doctrine of "manifestation"; suffice it to say that I discern no real conflict between the two positions. The point in either case is that the universe is ultimately to be explained in terms of a metacosmic reality, of which it is an effect or a partial manifestation. This implies, moreover, that such "being" as is to be found within the cosmos is indeed secondary or derived, a "being by participation" as Platonists say (which is also, presumably, the import of the *ego sum qui sum* of Exodus 3:14).

Now this perennial position is of course metaphysical in the literal sense of exceeding what physics is able to define or comprehend on the strength of its own proper methods; and yet it seems that the history of physics in our century can well be perceived as an indirect confirmation of that metaphysical doctrine. Let us recall, first of all, that besides the "orthodox" ontologies of its leading schools, antiquity has bequeathed to us also a "heterodox" ontology in the form of Democritean atomism; and it was this heterodox ontology, precisely, which in post-medieval times was reinstated by Descartes and soon imposed itself upon the educated thought of Western man. And so it went till about 1925 when the new quantum theory cast seemingly fatal doubts upon the conception of an ultimate particulate reality. The decisive event, however, came a few decades later with the discernment of ineluctable nonlocality, formulated first as a rigorous theorem of quantum mechanics, and later verified directly by certain remarkable experiments.

One could hardly ask for more: the erstwhile atomism has been deposed, though of course it lingers on as an almost irresistible ontological propensity or prejudice. But that is another matter, entirely. The point I wish to make is that the contemporary scientific refutation of Democritean atomism has opened the door to a serious reconsideration of the long-neglected ontology of the great traditions. I would say, in fact, that science today has need of this "perennial ontology" if it is to arrive not at a mere formalism that "works" but at a cogent interpretation of its own basic results.

3 What is your view on the origin of life: both on a scientific level and—if you see the need—on a metaphysical level?

I am opposed to Darwinism, or better said, to the transformist hypothesis as such, no matter what one takes to be the mechanism or cause (even perhaps teleological or theistic) of the postulated macroevolutionary leaps. I am convinced, moreover, that Darwinism (in whatever form) is not in fact a scientific theory, but a pseudo-metaphysical hypothesis decked out in scientific garb. In reality the theory derives its support not from empirical data or logical deductions of a scientific kind but from the circumstance that it happens to be the only doctrine of biological origins that can be conceived within the constricted *Weltanschauung* to which a majority of scientists no doubt subscribe. In other words, once the locus of reality has been narrowed to the categories of physics—and presumably, of Newtonian physics, no less—it is no longer possible to conceive of "speciation" otherwise than in basically Darwinist terms.

The matter stands differently, needless to say, if we are willing to take seriously the ontological conceptions of the premodern schools. For apart from the fact that "primary being" is assigned (according to all traditional schools) to a metacosmic plane, one needs also to recall that the cosmos itself is traditionally conceived as a hierarchy of distinct yet interpenetrating levels (a fact which presumably is not unrelated to Margenau's "transcendence with compatibility"). From this ontological point of vantage, moreover, our "physical universe" corresponds precisely to the outermost "shell", the plane of manifestation on which the "beings" that comprise our world attain a maximum of separation from each other as well as from their common ontological source.

Now this way of envisaging the cosmos reintroduces something that has been quite forgotten in Western thought since the demise of the Middle Ages, and that is the dimension of "verticality" and the possibility of "ontological transitions". In the enlarged perspective of traditional thought, life and the natural species originate, symbolically speaking, "at the center" and evolve (in the original sense of an "unfolding") towards the periphery: first comes a kind of spiritual seed (the *logos spermatikos* or *ratio seminale* of Western tradition); then comes an intermediate state of gestation; and finally a breakthrough into the domain of visible or "corporeal" manifestation. (As I have pointed out elsewhere, this "evolutive" process has been strikingly portrayed in the second chapter of Genesis [verses 4 and 50]. See *Teilhardism and the New Religion* [Rockford, IL: TAN Books, 1988], ch. 1)

James Gray once remarked (after commenting on the astronomical improbability of the Darwinist conjecture) that "most biologists feel it is better to think in terms of improbable events than not to think at all"; I would only add that the perennial and universal doctrine to which I have referred does at the very least deliver us from this predicament.

4 What is your view on the origin of *Homo sapiens*?

Having rejected the transformist hypothesis I have also of course denied the notion that man has descended from subhuman stock. I would add that man is special, not in the mode of his origination, but quite simply in what he is, in the archetype, namely, which he manifests. And on this point all religions and all sapiential traditions are in perfect agreement: man is the theomorphic creature *par excellence*, whence his preeminence and his position of centrality.

Perhaps the most baneful consequence of evolutionist thought is that it obscures, more effectively than any other pseudo-philosophy, the true nature of man and the loftiness of his destiny. One cannot but agree with Seyyed Hossein Nasr when he writes (with reference to the Darwinist age in which we live) that "Never before has there been so little knowledge of man, of the *anthropos*".

The fact is that Darwinism constitutes a counterposition to the perennial wisdom of mankind. It represents a systematic denial of the archetypes, of the essences, of that "participation in Being" upon which all life and all existences hinge. In the climate of Darwinist thought most of what the religions teach loses its meaning or, worse still, assumes another and indeed contrary sense. To be sure, there have been attempts to fuse evolutionism and religion; but the point is that these new interpretations of perennial doctrines have in fact falsified and corrupted what they pretend somehow to restore or render more palatable to the contemporary taste. Teilhard de Chardin, for instance, has unquestionably falsified Christianity even as Sri Aurobindo has mutilated Hinduism.

At bottom evolutionism is the denial of transcendence, the desperate attempt to understand life on the horizontal plane of its manifestations. Religion, on the other hand, is perforce concerned with transcendence and the vertical dimension, in which alone the *re-ligare* or "binding back" can be effected. The supposed merger, therefore, of these opposed doctrines constitutes one of the most bizarre happenings in these already confused and confusing times.

5 How should science—and the scientist—approach origin questions, specifically the origin of the universe and the origin of life?

If one takes seriously the traditional ontologies (as I believe one should), it becomes evident that the problem of origins is inherently metaphysical, simply because the origins in question involve a transition between distinct ontological planes. We know of an impressive book entitled *The First Three Minutes*; it does not tell us, however, what exists or transpires at $t = 0$. It obviously cannot, not because nothing then exists, but because the reality in question is not subject to the conditions or categories in terms of which our scientific descriptions are necessarily framed. Metaphysically speaking, origins are never situated on the posterior ontological plane.

There are of course "origins" of the different kind—the formation of a molecule from pre-existent atoms, for instance—to which the preceding observations do not apply; we might refer to the latter as "origins of the second kind" to distinguish them from the former class, the "primary origins" if you will. There is presumably a primary origin "at time t = 0"; we would like however to suggest that primary origins are legion. (In a sense, perhaps, all primary origins take place "at time t = 0", for there is much to be said for the view that "outside" the physical univese one finds oneself always "at the beginning". On this subject we refer to *Cosmos and Transcendence* ([Peru, IL: Sherwood Sugden and Company, 1984], chapter 3). There are necessarily innumerable primary origins simply because the physical universe is not in fact the closed and "self-sufficient" domain we have taken it to be. There must also, moreover, be "ultimate endings"; which is to say that here is a kind of "two-way commerce" between the physical and the higher ontological planes. Fantastic as it may sound to a "Newtonian mentality", there is a Jacob's Ladder, perhaps, along which beings "ascend and descend perpetually".

Science, one can say, is constrained to deal with things that have already "originated"; it deals, in other words, with things that exist on the physical plane. There was a time, not long ago, when this statement would have been received as an unmitigated truism; but times are changing. From a metaphysical standpoint, in any case, the statement, so far from being a truism, expresses in fact a most stringent limitation of the scientific enterprise. What it signifies, quite clearly, is that science, by the very nature of its methods, is debarred from grasping primary origins and ultimate terminations; and I would add that this categorical limitation is especially restrictive in the biosphere, where birth and death abound.

How, then, should the scientist "approach origin questions"? With due modesty, I would say, born of the sobering recognition that his methods provide access only to a certain "outer shell" of things. There is of course more than enough work to do for the scientists within his own proper domain; and it could also be said that, to him who has "eyes to see", that domain, bounded though it be, points beyond itself—even to "the invisible things of God". At the very least, however, the scientist should be aware of the distinction between primary and secondary origins and wary of forcing the former into the mold of the latter as the Darwinists have done. One should remember that science turns forthwith into pseudo-science and superstition the moment it oversteps its own proper bounds.

6 Many prominent scientists—including Darwin, Einstein, and Planck—have considered the concept of God very seriously. What are your thoughts on the concept of God and on the existence of God?

To me personally nothing is more evident, more certain, than the existence or reality of God. I incline in fact to the view that the existence of God constitutes indeed the only absolute certainty, even as he (or it) constitutes, in the final analysis, the only true or absolute Existent (in conformity with the Vedantic teaching or the *ego sum qui sum* of Judeo-Christian revelation).

However, it needs also to be recognized that belief in God is subject to degrees; and few there are, one must admit, whose belief in God is altogether whole or unimpaired. One has every right to surmise, moreover, that belief in God, when it *is* unimpaired, is invariably accompanied by a permanent realization of God in all that exists.

Since God is the source of all being, it behooves us to "believe" in God with all that we are; which is to say that belief in God, when it is whole, involves not only the mind, but necessarily every part or faculty of the human constitution ("Thou shalt love the Lord thy God with all thy heart, and with all thy soul, and with all thy mind, and with all thy strength"). Pascal was right, therefore, when he distinguished between "the God of the philosophers" and "the living God of Abraham": between the God in whom one believes "with a corner of one's mind" and the God of religion, the God of the saints.

It is easy to forget in this nominalistic age that to believe in God is to "participate" in Him in a corresponding degree; and this implies that religious faith, when it is the real thing, is not at all the blind acceptance of unprovable dogmas it is so often made out to be but a certain participation in realities of a supra-sensible order. As such, moreover, religious faith stands head and shoulders above every form of knowledge—no matter how scientific or exact—that does not reach beyond the phenomenal plane.

Finally, what bearing, if any, do these considerations have upon the "problem of origins"? Simply this: to resolve the riddle

of origins in truth is ultimately to know the one Origin from which everything in the universe has sprung—and that is God. To be sure, scientific explanations have their validity, their own fascination and use; they do not, however, resolve the problem but only shift the enigma to a deeper plane. If science begins in wonder, as indeed it does, it ends perforce in a sense of wonder that is greater still. Thus, to give a particularly striking example, it is the latest physics, precisely, that inspires in us a profound sense of wonder and awe in the face of the physical universe; how paltry by comparison was the "Newtonian model", and how unspeakably naive the "experts" who mistook that model for the Reality itself. If the physics of the last century has prompted atheism, the physics of today is inciting at least the more thoughtful of its votaries to reexamine "the question of God".

24

The Guidance of Evolution Lets God Appear to Us in Many Guises

• • • • • • • • • •

Professor Walter Thirring

- Born 29 April 1927

- Ph.D. in theoretical physics, University of Vienna, 1949

- Currently Director of the Institute for Theoretical Physics and Professor at the University of Vienna

- Areas of specialization and accomplishments: research on theory of elementary particles, theory of gravitation, solid state physics; works include *Einführung in die Quantenelektrodynamik,* 1954; *Principles of Quantum Electrodynamics* (with E.M. Henley), 1958; *Elementary Quantum Field Theory,* 1962

• • • • • • • • • •

These fundamental questions of mankind require a more extensive analysis. Therefore excuse me if I enlarge only on Question 6 (on the existence of God). This may also shed some light on my opinion on the other points.

I think that scientists who devote their lives to exploring the *harmonia mundi* cannot help seeing in it some divine plan. So the question is not so much whether they believe in the existence of God but what kind of notions they connect with this word and with what attributes they would like to endow him. As H. Bondi once said, "atheism is logically not sound since how can one deny the existence of something which is only so vaguely defined". Regarding what and how God is, there are a wide range of attitudes; some scientists (Einstein, Planck) did not believe in a personal God, others (Pascal) did. If we agree to talk about God the creator then he appears to the cosmologist as an abstract law which governed the Big Bang, far remote from any human feature. To the extent that these laws determine uniquely the evolution afterwards he cannot acquire any more personal aspects. He expresses himself only in the *Urgleichung* ("basic equation"). I do not share this view; to me the laws of nature appear of another

119

structure. They form an infinite hierarchy where each lower level is not in contradiction with the upper level but is not completely determined by it either. In fact the essential features of the lower level appear accidental when looked at from above.

To explain what I mean let me for the moment assume that there is some justice in the present view that the original law of the universe was some supersymmetric field theory which in the course of evolution, by some phase transition, broke down to what we see now, namely strong, electroweak, and gravitational interactions. The functioning of our universe depends sensitively on the strength of these interactions, but from the point of view of the original theory these parameters are of accidental nature, as unpredictable as the thickness of the ice on a lake after a cold night.

These highly speculative theories may not convince, but in the past decades it has become increasingly clear that the complex systems we are surrounded with are intrinsically random in the sense that any uncertainty grows exponentially in time. Since in quantum theory these uncertainties are unavoidable, the classical determinism is an illusion. Laplace's answer "je n'ai pas besoin de cette hypothèse" when Napoleon asked him question number 6 only shows that at that time the compexity of dynamical systems was not appreciated. Today all these accidental features of evolution are put into the Anthropic Principle which says that they always turn in such a way that eventually man can live. It is as if God were continuously guiding the evolution. In such situations one eventually drops the "as if", and I shall do so in the sequel. This guidance of the evolution lets God appear to us in many guises: as the fundamental law which governed the Big Bang to the cosmologist; as the chance which determined the fate of people to the historian; as the one who revealed himself in the prophets and in Jesus Christ to the theologian. It appears to me as if God himself had evolved together with the cosmos and became more personal as it developed from a hot plasma cloud to a highly structured state containing beings like us.

Some scientists consider mankind only as an accidental feature of the universe since they fail to see us imprinted in what they think is the fundamental law of physics. Thus they refuse to give us the title of "coronation of the creation" and do not believe that God guides us individually since he is much too busy with bigger astrophysical jobs. Such thoughts are foreign to my mind as I do not believe that I can understand God with my human logic. I can only appeal to my personal experience when I believe that he guides me as he appears to do with every little bit of his creation.

The fierce battles between scientists and theologians seem to me not so much inherent to these subjects but rather due to the pretentious character of some of their representatives who believe that they understand more than they do. This becomes better only once one has learned the due humility vis-à-vis the great mysteries of the cosmos.

25

The Question of Origin Seems Unanswered if We Explore from a Scientific View Alone
· · · · · · · · · ·
Professor Charles H. Townes

- Born 28 July 1915

- Ph.D. in physics, California Institute of Technology, 1939; Nobel Prize for Physics (shared with Nikolai Basov and Aleksandr Prokhorov), 1964; received the Nobel Prize with Basov and Prokhorov "for [their] fundamental work in the field of quantum electronics, which has led to the construction of oscillators and amplifiers based on the maser-laser principle"

- Currently University Professor of Physics at the University of California, Berkeley

- Works include *Molecular Microwave Spectra Tables* (with Paul Kisliuk), 1952; *Microwave Spectroscopy* (with Arthur L. Schawlow), 1955

- Professor Townes on:

   the origin of the universe: "It is true that physicists hope to look behind the 'Big Bang' and possibly to explain the origin of our universe as, for example, a type of fluctuation. But then, of what is it a fluctuation and how did this in turn begin to exist?"

   the origin of life: ". . . I do not know how life originated".

   the origin of *Homo sapiens:* ". . . mankind developed in the more or less accepted evolutionary way from early forms of life. . . . we do not understand this evolution very well—for example, the apparently large evolutionary jump which led to man's facility with language, mathematics, and related ideas".

   God: "I believe in the concept of God and in his existence".

· · · · · · · · · ·

1 What do you think should be the relationship between religion and science?

   I regard religion and science as two somewhat different approaches to the same problem, namely that of understanding ourselves and our universe. To that might be added, in the case of religion, understanding of the purpose of our universe. However, I would not exclude this from science either. Thus, in my view, religion and science are aimed at much the same problem and must in time converge.

2 What is your view on the origin of the universe: both on a scientific level and—if you see the need—on a metaphysical level?

I do not understand how the scientific approach alone, as separated from a religious approach, can explain an origin of all things. It is true that physicists hope to look behind the "Big Bang" and possibly to explain the origin of our universe as, for example, a type of fluctuation. But then, of what is it a fluctuation and how did this in turn begin to exist? In my view, the question of origin seems always left unanswered if we explore from a scientific view alone. Thus, I believe there is a need for some religious or metaphysical explanation if we are to have one.

3 What is your view on the origin of life: both on a scientific level and—if you see the need—on a metaphysical level?

My view of the origin of life is not dissimilar from the ordinary scientific one. That is, I believe that life originated from a concatenation of molecular reactions which somehow produced self-reproducing systems and eventually evolved to our present variety of life forms. Nevertheless, I am not at all sure that our present scientific understanding is adequate to explain such a development. Very likely new ideas about complex systems and interactions will be needed. This may even require some type of organizing force or principles which we do not presently recognize. The simple statement at this point is that I do not know how life originated.

4 What is your view on the origin of Homo sapiens?

I believe that mankind developed in the more or less accepted evolutionary way from early forms of life. Again, we do not understand this evolution very well—for example, the apparently large evolutionary jump which led to man's facility with language, mathematics, and related ideas. As in the case of the origin of life, one can wonder whether some new principles are involved. However, the necessity for new principles may be somewhat less persuasive in this case than in the origin of life itself.

5 How should science—and the scientist—approach origin questions, specifically the origin of the universe and the origin of life?

> I do not know of any special and unique approaches that the scientists should use towards understanding the origin of things. Scientific methods are already quite diverse; we use every technique we can think of and should continue to do so, inventing new ones as possible. To what extent that will give us a substantially deeper answer to the origin of the universe is not clear to me. However, I do believe that scientific methods and their extensions by human wits will likely help us understand the origin of life in the long run.

6 Many prominent scientists—including Darwin, Einstein, and Planck—have considered the concept of God very seriously. What are your thoughts on the concept of God and on the existence of God?

> I believe in the concept of God and in his existence.

26

The Origin of the Universe Can Be Described Scientifically as a Miracle

· · · · · · · · · ·

Professor Herbert Uhlig

- Born 3 March 1907

- Ph.D. in chemistry, Massachusetts Institute of Technology, 1932; he has been awarded professional honors by the Electrochemical Society, the Institute of Corrosion Science and Technology of Great Britain, and the National Association of Corrosion Engineers; the Corrosion Laboratory at the Massachusetts Institute of Technology is named in his honor

- Professor Emeritus in the Department of Materials Science and Engineering, Massachusetts Institute of Technology

- Areas of specialization and accomplishments: research on corrosion and oxidation of metals, nature and source of corrosion resistance exhibited by stainless steels and other corrosion resistant metals and alloys, effect of chemical environments on fracture of metals, metallurgical evidence regarding origin of meteorites; works include *Corrosion and Corrosion Control*, 1963

- Professor Uhlig on:

    the origin of the universe: "The origin of the universe can be described scientifically as a miracle. A scientific miracle is here defined as a natural event having a very small probability of happening".

    the origin of life: Can be similarly described as a miracle.

    the origin of *Homo sapiens:* Evolution of the original simple living cell or cells.

    God: "Faith in the concept of a God . . . is essential to . . . ultimate survival of the human race".

· · · · · · · · · ·

1 What do you think should be the relationship between religion and science?

Both represent the truth. Science, being relatively recent to human experience, is gradually approaching, by different routes, the conclusions reached previously through religous sources.

2 What is your view on the origin of the universe: both on a scientific level and—if you see the need—on a metaphysical level?

> The origin of the universe can be described scientifically as a miracle. A scientific miracle is here defined as a natural event having a very small probability of happening.

3 What is your view on the origin of life: both on a scientific level and—if you see the need—on a metaphysical level?

> Life, having a singular non-repetitive first appearance over the past several billion years, has an origin similarly described as a miracle. (See my *Life, Science and Religious Concerns*, pp. 33–42.)

4 What is your view on the origin of *Homo sapiens?*

> The origin of *Homo sapiens* is logically described in terms of evolution of the original simple living cell or cells.

5 How should science—and the scientist—approach origin questions, specifically the origin of the universe and the origin of life?

> Proposed origins, being speculative at best, justifiably take into account scientific conclusions based on current observations and available facts, but the inherent uncertainty of any such extrapolations should always be acknowledged.

6 Many prominent scientists—including Darwin, Einstein, and Planck—have considered the concept of God very seriously. What are your thoughts on the concept of God and on the existence of God?

> Faith in the concept of a God who is concerned with his creation is essential to human hope, an optimistic world view, and ultimate survival of the human race. Any contrary view aligns humanity with the frustration of a drifting, meaningless universe facing a despondent future.

27

There Is a Bohr Complementarity between Science and Religion

• • • • • • • • • •

Professor Victor Weisskopf

• Born 19 September 1908

• Ph.D. in physics, University of Göttingen, 1937; received the Max Planck medal, 1956, and the U.S. Department of Energy's 1988 Enrico Fermi Award for outstanding scientific or technological achievement in the development, use, or control of atomic energy

• Areas of specialization and accomplishments: research on theories of elementary particles, nuclear phenomena, quantum dynamics and electrodynamics, electron theory, nuclear physics; works include *The Privilege of Being a Physicist*, 1989

• • • • • • • • • •

To answer them (the six questions) one would have to write several long papers which I am unable to do now. I am finishing my autobiography, in which some of the questions are treated. Also my recent essay collection, *The Privilege of Being a Physicist*, contains some ideas of that type, especially the essays "Art and Science" and "The Frontiers and Limits of Science". As to the origin of the universe, look at my article in the February 16 (1989) issue of *New York Review of Books*. Also in the January 1989 issue of the *Bulletin of the American Academy of Arts and Sciences*.

God?!

In a Jewish theological seminar there was an hours-long discussion about proofs of the existence of God. After some hours, one rabbi got up and said: "God is so great, he does not even need to exist". (Existing is a category, not applicable to God, only to you, me, the table, and so on.) There is a Bohr complementarity between science and religion.

28

The Origin of Time Is Not in Time

· · · · · · · · · ·

Professor C. F. von Weizsacker

- Born 28 June 1912

- Ph.D. in physics (with Werner Heisenberg), University of Leipzig, 1933; received the Max Planck Medaille, 1958, and the Templeton Prize, 1989; delivered the Gifford Lectures ("The Relevance of Science"), 1949

- Director of the Max Planck Institute on the Preconditions of Human Life in the Modern World, Starnberg

- Areas of specialization and accomplishments: research on astrophysics and cosmology, theory of the origin of the solar system, galactic systems, and evolution of stars; studies in atomic physics, including axiomatic foundation of quantum theory, quantum logic, and a unified theory of elementary particle physics with cosmology; works include *Die Atomkerne: Atomenergie und Atomzeitalter,* 1958; *Zum Weltbild der Physik,* 1963; *Die Tragweite der Wissenschaft,* 1964; *Aufbau der Physik,* 1985; *Bewusstseinswandel,* 1988

- Professor von Weizsacker on:

   the origin of the universe: "The word *origin* is not clear. If it means origin in time the main question is not asked, which is, what is the origin of time itself? The origin of time is not in time".

   the origin of life: "I am quite satisfied with the origin of life as described by modern theories of molecular biology. The question then is, what are molecules?"

· · · · · · · · · ·

1 What do you think should be the relationship between religion and science?

I think the relationship between religion and science should be friendship. In the last resort their messages might turn out to be identical, but this would mean both of them would have to mature. At present I feel that just in friendship they should ask each other critical questions. Religion might ask science whether scientists understand the immense danger into which they bring the world, mankind, and the living world on our earth, if they do not take their responsibility for the future as their first duty. On the other hand, science should ask religion, in all friendship, whether it is not resting on concepts which are outmoded by five hundred years or more. Both of them ought to mature and then they would be far closer to each other than they realize today.

2 What is your view on the origin of the universe: both on a scientific level and—if you see the need—on a metaphysical level?

The word *origin* is not clear. If it means origin in time the main question is not asked, which is, what is the origin of time itself? The origin of time is not in time. Martin Luther once said, when he was asked what God did in the very long time before he created the world: "He sat in a birch grove cutting whips for people who asked unnecessary questions". In fact, I feel that the Big Bang may be quite a good approximate hypothesis, but if you believe too strictly in it I never know whether this is not just the creation myth of the very same century in which nuclear weapons were invented. In general, our mythical ideas express our basic feelings. Scientists in general and many theologians too have no suffcient training in philosophy. I think the description of all these things in a philosophy like that of Plato is on a much higher conceptual level than nearly all the debates I hear in our days. Still Plato does not know what we know about what we call "the open future", and hence I cannot just accept what he says. He himself would have laughed at somebody who would think two thousand years after him that he could accept it. But on the other hand, the level of investigation ought to be understood.

3 What is your view on the origin of life: both on a scientific level and—if you see the need—on a metaphysical level?

Scientifically I am quite satisifed with the origin of life as described by modern theories of molecular biology. The question then is, what are molecules? Quantum theory, I think, would be fully in agreement with the idea of what I call a spiritualistic monism, that is, that there is not something which is not living and beside which there is life which ought to be explained, but that the basic essence of the universe is to live. Only again these are words, and the important thing is to have a good philosophy using these words. This cannot be done in answering a few questions.

4 What is your view on the origin of *Homo sapiens*?

> I once had a little monkey for twelve years in my home, and he was my very best argument for Darwin, because to be descended from such a nice being is just wonderful.

5 How should science—and the scientist—approach origin questions, specifically the origin of the universe and the origin of life?

> I think science and the scientists should approach such questions by asking what they mean.

6 Many prominent scientists—including Darwin, Einstein, and Planck—have considered the concept of God very seriously. What are your thoughts on the concept of God and on the existence of God?

> The concept of God is quite a nice thing and very difficult to discuss, but I have known some great scientists myself, and it was absolutely clear that essentially they believed in the inexplicable, which people who don't have a better word describe by the idea of God.

29

The Origin of the Universe Is a Disturbing Mystery for Science

· · · · · · · · · · ·

Professor Eugene P. Wigner

- Born 17 November 1902

- Dr. Ing. in chemical engineering, Technische Hochschule, Berlin, 1925; Nobel Prize for Physics (shared with J. Hans D. Jensen and Maria Goeppert Mayer), 1963; received the Nobel Prize "for his contributions to the theory of the atomic nucleus and the elementary particles, particularly through the discovery and application of fundamental symmetry principles"

- Emeritus Professor of Physics, Princeton University

- Works include *Nuclear Structure* (with Leonard Eisenbud), 1958; *The Growth of Science: Its Promise and Its Dangers*, 1964; *Symmetries and Reflections: Scientific Essays*, 1967

- Professor Wigner on:

    the origin of the universe: "The origin of the universe is a mystery for science, surely for the present one. It is a disturbing mystery".

    the origin of life and of *Homo sapiens*: "The same applies. . ."

    God: "The concept of God . . . helps us to make decisions in the right direction. We should be very different, I fear, if we did not have that concept."

· · · · · · · · · ·

1 What do you think should be the relationship between religion and science?

Science presents very attractive purposes. It is natural for man to try to add to it. Religion serves mainly as a directive. They interact but very little.

2 What is your view on the origin of the universe: both on a scientific level and—if you see the need—on a metaphysical level?

The origin of the universe is a mystery for science, surely for the present. It is a disturbing mystery.

3 What is your view on the origin of life: both on a scientific level and—if you see the need—on a metaphysical level?

4 What is your view on the origin of *Homo sapiens*?

> The same answer applies here as to your Questions 2.

5 How should science—and the scientist—approach origin questions, specifically the origin of the universe and the origin of life?

> He should, perhaps but only perhaps, devote thought to it. But it may remain an unsolved question for us.

6 Many prominent scientists—including Darwin, Einstein, and Planck—have considered the concept of God very seriously. What are your thoughts on the concept of God and on the existence of God?

> The concept of God is a wonderful one—it also helps us to make decisions in the right direction. We would be very different, I fear, if we did not have that concept. We should abide by it even though present day science does not support it.

30

The Hidden Variables of Quantum Mechanics Are Under God's Power

• • • • • • • • • •

Professor Shoichi Yoshikawa

- Born 9 April 1935

- Ph.D. in nuclear engineering, Massachusetts Institute of Technology, 1961; received the Mainichi Publication Award in 1975 for his work *Challenge to Nuclear Fusion*

- Senior Research Scientist and Professor, Department of Astrophysical Sciences, Princeton University

- Areas of specialization and accomplishments: experimental and theoretical plasma physics, nuclear fusion device design; works include *Introduction to Controlled Thermodynamic Research* (with A. Iiyoso), 1972; *Challenge to Nuclear Fusion,* 1974

- Professor Yoshikawa on:

  the origin of the universe: "I agree with the basic points of the . . . so-called 'Big Bang' theory. . . ."

  the origin of life: "I believe the [Darwinian] theory explains some of the cases of evolution," but all varieties of DNA chains and egg cells cannot be accounted for "by the mechanical processes only".

  the origin of *Homo sapiens:* ". . . may not be understood completely".

  God: "I think that God originated the universe and life".

• • • • • • • • • •

1 What do you think should be the relationship between religion and science?

Personally, I subscribe to the old notion that (monotheistic) religion is above science. However, in my everyday conduct, I follow the general norm of present-day society that religion and science have different domains of influence on the human being.

2 What is your view on the origin of the universe: both on a scientific level and—if you see the need—on a metaphysical level?

I agree with basic points of the current theory of the birth of the universe (the so-called "Big Bang" theory). Observational evidences (homogeneity of the universe, the remnant radiation, deuterium/hydrogen ratio, and so on) overwhelmingly support this view. Of course, I expect some modifications of the current theory in the future.

I see no contradiction with the description of the Old and New Testament concerning the birth of the universe.

3 What is your view on the origin of life: both on a scientific level and—if you see the need—on a metaphysical level?

I understand the Darwinian theory, and I believe the theory explains some of the cases of evolution where the external conditions remain static. For example, the development of organisms in an isolated lake can be understood within the framework of the theory of evolution. However, it is very difficult for me to believe that all the evolution or, more precisely, the existence of all the varieties of DNA chains and egg cells can be accounted for by the mechanical processes only.

4 What is your view on the origin of *Homo sapiens*?

In particular, the origin of *Homo sapiens* may not be understood completely. The seemingly random events of nature, such as glacier periods, the apparent size of the moon versus the sun, volcanic eruptions, the existence of microorganisms (viruses, bacteria, and so forth) have influenced the development of the human being. Were they just coincidences? Also the concept of beauty (music, poetry, paintings, and so on) appears to be shared by many. The mechanical explanations could be advanced, but . . .

5 How should science—and the scientist—approach origin questions, specifically the origin of the universe and the origin of life?

The origin of the universe can be understood on a scientific level. It does not conflict with religion. The origin of life could be explained by the theory of evolution. However, it is equally possible that, during the birth of a new life, the selection of a new set of genes is somehow influenced by a metaphysical force.

6 Many prominent scientists—including Darwin, Einstein, and Planck—have considered the concept of God very seriously. What are your thoughts on the concept of God and on the existence of God?

I think that God originated the universe and life. *Homo sapiens* was created by God using the process that does not violate the physical laws of the universe significantly or none at all. (Hidden varibles of quantum mechanics under God's power?)

PART TWO

. . . .

Biologists and Chemists

1

There Exists an Incomprehensible Power with Limitless Foresight and Knowledge

· · · · · · · · · ·

Professor Christian B. Anfinsen

- Born 26 March 1916

- Ph.D. in biochemistry, Harvard University, 1943; Nobel Prize for Chemistry (shared with Stanford Moore and William H. Stein), 1972; received the Nobel Prize "for his work on ribonuclease, especially concerning the connection between the amino acid sequence and the biologically active conformation"

- Currently Professor of Biology at Johns Hopkins University

- Works include *The Molecular Basis of Evolution*, 1959

- Professor Anfinsen on:

    the origin of the universe: "I cannot help but believe the conclusions that grow out of . . . work on the Big Bang".

    the origin of life: ". . . an inevitable consequence of the evolution of the universe, physically speaking".

    the origin of *Homo sapiens:* ". . . developed from lower forms by . . . processes of mutation and selection".

    God: "I think only an idiot can be an atheist. We must admit that there exists an incomprehensible power or force with limitless foresight and knowledge that started the whole universe going in the first place".

· · · · · · · · · ·

1 What do you think should be the relationship between religion and science?

> Religion—or I should say, the religions of the world—are a natural outgrowth of man's need for some answer, however mystical, to his concerns about existence, behavior, and morality. Science is quite a separate subject and deals with man's eternal curiosity about how the world and the universe are constructed and the laws that operate physically.

2 What is your view on the origin of the universe: both on a scientific level and—if you see the need—on a metaphysical level?

> I cannot help but believe the conclusions that grow out of the Penzias-Wilson experiments and other work on the Big Bang. Since we exist (I presume) there must have been a beginning and, scientifically, this appears to have occurred something like twenty billion years ago. The moment of the Big Bang included the built-in galaxy of physical laws that have inevitably led to the universe as we see it now.

3 What is your view on the origin of life: both on a scientific level and—if you see the need—on a metaphysical level?

> The origin of life, it seems to me, was an inevitable consequence of the evolution of the universe, physically speaking. There came a time when a combination of elements, heat, water, and who knows what else, led to the formation of a living thing—that is, an object that could reproduce and could be susceptible to mutation and selection in the Darwinian sense.

4 What is your view on the origin of *Homo sapiens*?

> Like all other living things, *Homo sapiens*, in my view, developed from lower forms by the generally accepted processes of mutation and selection. Indeed, even the Vatican appears to be happy about man's descent from lower forms of life.

5 How should science—and the scientist—approach origin questions, specifically the origin of the universe and the origin of life?

> Science and the scientist should continue to approach the questions of origin as is now being done by studying more and more primitive organisms such as the archae bacteria and other forms of life that perhaps resemble the original reproducing forms.

6 Many prominent scientists—including Darwin, Einstein, and Planck—have considered the concept of God very seriously. What are your thoughts on the concept of God and on the existence of God?

> I think only an idiot can be an atheist. We must admit that there exists an incomprehensible power or force with limitless foresight and knowledge that started the whole universe going in the first place. Such a process may have occurred many times earlier, and, indeed, must have, and will very likely occur again in the

future. I enclose a favorite quotation from Einstein that agrees almost completely with my own point of view.

Einstein himself once said that "the most beautiful and most profound emotion we can experience is the sensation of the mystical. It is the sower of all true science. He to whom this emotion is a stranger, who can no longer stand rapt in awe is as good as dead. That deeply emotional conviction of the presence of a superior reasoning power, which is revealed in the incomprehensible Universe, forms my idea of God."

2

The Existence of a Creator Represents a Satisfactory Solution

• • • • • • • • • •

Professor Werner Arber

• Born 23 June 1929

• Ph.D. in biophysics, University of Geneva, 1958; Nobel Prize for Physiology/Medicine (shared with Daniel Nathans and Hamilton O. Smith), 1978; received the Nobel Prize for "the discovery of restriction enzymes and their application to problems of molecular genetics"

• Currently Professor of Microbiology at the University of Basel

• Professor Arber on:

the origin of the universe: ". . . today's universe is the result of a steady evolution".

the origin of life: ". . . life only starts at the level of a functional cell. . . . How such already quite complex structures may have come together, remains a mystery to me. The possibility of the existence of a Creator, of God, represents to me a satisfactory solution to this problem."

the origin of *Homo sapiens:* Man is "a living organism as any other" with "well-developed intellectual abilities . . . that we do not fully understand yet".

God: "I range the concept of God among . . . the principles acting in the universe".

• • • • • • • • •

1 What do you think should be the relationship between religion and science?

I see religion as an important help to man in coping with daily questions in respect to his life in society. Certain types of ideology may take the place of a religion but science cannot, although some people tend to claim that it does. Religion and science belong not to the same category of influences on man. Religions often use parascientific descriptions such as the description of creation. However, this is not the principle goal of religion; religion only uses the description in order to illustrate its basic principles. For this reason a religion should be flexible enough to adapt to increasing knowledge and to incorporate the results of scientific advance. Only thus can a religion serve its purpose to humans.

2 What is your view on the origin of the universe: both on a scientific level and—if you see the need—on a metaphysical level?

As to the origin of the universe, my personal solution is relatively simple. I assume that in cosmic times the time-scale is not linear; in this way, I can push back the origin of the universe to infinity. This may be a simple-minded view, but it liberates me from worries about not understanding how everything came about. In this view, today's universe is the result of a steady evolution.

3 What is your view on the origin of life: both on a scientific level and—if you see the need—on a metaphysical level?

Although a biologist, I must confess that I do not understand how life came about. Of course, it depends on the definition of life. To me, autoreplication of a macromolecule does not yet represent life. Even a viral particle is not a life organism, it only can participate in life processes when it succeeds in becoming part of a living host cell. Therefore, I consider that life only starts at the level of a functional cell. The most primitive cells may require at least several hundred different specific biological macro-molecules. How such already quite complex structures may have come together, remains a mystery to me. The possibility of the existence of a Creator, of God, represents to me a satisfactory solution to this problem.

4 What is your view on the origin of *Homo sapiens*?

Being interested myself in better understanding the mechanism of biological evolution, I do not have problems understanding the origin of *Homo sapiens*. Biologically, man is just a living organism as any other. Of course, he has well-developed intellectual abilities, and we do not fully understand yet how this came about. But there is no good scientific evidence to assume that *H. sapiens* is an independent creation. Taking this argument further, man should realize that he is stringently dependent in his life on a multitude of other living organisms, as any other living organisms are. Man therefore should stop drastically interfering with ecological equilibria in such an intensive manner that further biological evolution is endangered because of a serious reduction in biological diversity.

5 How should science—and the scientist—approach origin questions, specifically the origin of the universe and the origin of life?

As may be obvious from my answers to Questions 2 and 3, I have no useful answer to this question.

6 Many prominent scientists—including Darwin, Einstein, and Planck—have considered the concept of God very seriously. What are your thoughts on the concept of God and on the existence of God?

I do not think that our civilization has succeeded in discovering and explaining all the principles acting in the universe. I include the concept of God among these principles. I am happy to accept the concept without trying to define it precisely. I know that the concept of God helped me to master many questions in life; it guides me in critical situations, and I see it confirmed in many deep insights into the beauty of the functioning of the living world.

3

The Ultimate Truth Is God

· · · · · · · · · ·

Professor D. H. R. Barton

- Born 8 September 1918

- Ph.D. in organic chemistry, Imperial College of Science and Technology, University of London, 1942; Nobel Prize for Chemistry (shared with Odd Hassel), 1969; received the Nobel Prize with Hassel "for their contributions to the development of the concept of conformation and its application in chemistry"

- Currently Professor of Chemistry at Texas A&M University and Director of the Institute for the Chemistry of Natural Substances in Gif-sur-Yvette, France

- Works include *Some Modern Trends in Organic Chemistry*, 1958; *Specific Fluorination in the Synthesis of Biologically Active Compounds*, 1969

- Professor Barton on:

  the origin of the universe: ". . . the universe that we observe today [once] was compressed into a point which exploded".

  the origin of life: "Molecular biology is rapidly making clear the chemical principles on which life is based".

  **the origin of *Homo sapiens:* "*Homo sapiens* is a logical product of the evolutionary process".

  God: "God is Truth. . . . God shows himself by allowing man to establish truth".

· · · · · · · · · ·

1 What do you think should be the relationship between religion and science?

God is Truth. There is no incompatibility between science and religion. Both are seeking the same truth. Science shows that God exists. Our universe is infinitely large and infinitely small. It is infinite in time past and in future time. We can never understand infinity. It is the ultimate truth, which is God.

When scientists can make numerous repeatable experiments or observations, they establish truth. Religion seldom wishes to make experiments and the truth that is accepted is often divine intervention of God in the affairs of man. However, what is written is written by man, with the liability of human frailty. The observations and experiments of science are so wonderful that the truth that they establish can surely be accepted as another manifestation of God. God shows himself by allowing man to establish truth.

2 What is your view on the origin of the universe: both on a scientific level and—if you see the need—on a metaphysical level?

There is evidence that at one point in time the universe that we observe today was compressed into a point which exploded. Why not? But the matter of the universe had an infinite existence before this happened and will have an infinite future. God may well choose to redistribute matter and energy from time to time.

3 What is your view on the origin of life: both on a scientific level and—if you see the need—on a metaphysical level?

The origin of life is much easier to understand. Life began at a definite point in time and has evolved steadily since. What is life? Life is an association of molecules which can reproduce itself and which can evolve in response to external stress. At first our world, as it cooled, contained molecules in the seas based on carbon, hydrogen, oxygen, and nitrogen. There was a reducing atmosphere and hence the oxygen in these primitive molecules came from water. The simple molecule hydrogen cyanide was surely the most important. It was naturally abundant and from its polymers can be derived the four bases of the code of life. Professor Albert Eschenmoser (E.T.H. Zurich) has recently made a brilliant analysis of this theory. Phosphoric acid, to make phosphates, was another abundant constituent of the seas. Ribonucleic acid is a polymer of the four bases, phosphoric acid, and the sugar D-ribose. Optically inactive sugars can also be made anaerobically by the polymerization of formaldehyde, itself formed from reduction of hydrogen cyanide. Amino acids and hence peptides and proteins are also formed without oxygen and under reducing conditions.

The stage was now set for life to begin. Perhaps it only began once because all life today (except perhaps the mysterious "scrapie" virus) is related. The same genetic code of four bases is always used and the sugar is always 2-deoxy-D-ribose. Joining the four bases to the sugar and making a polymer using phosphoric acid gives us the genome of life—deoxyribonucleic acid (DNA). It is probable that life first used a similar polymer of ribonucleic acid (RNA) based on D-ribose itself. DNA is used by present-day life-forms because it is more stable than RNA and thus ensures a more faithful reproduction of the genome. However, even now, to express the genome as protein and make the catalysts of life (the enzymes) the DNA is retransformed into RNA.

Why life chose D-ribose is not clear. It is known today that derivatives of this sugar are first converted into 2-deoxy-D-ribose derivatives during the asembly of the DNA. All this seems to an organic chemist to be a clumsy way to go about the life process. But it works very efficiently. The genome is reproduced very faithfully and there are enzymes which repair the DNA, where errors have been made or when the DNA is damaged.

Molecular biology is rapidly making clear the chemical principles on which life is based. Using techniques from enzymology and organic synthesis, whole genes can be made quite easily. These genes can be expressed using the protein machinery of simple organisms like *E. coli*. We are only at the beginning of gene synthesis. In the future there will be difficult moral problems for man.

4 What is your view on the origin of *Homo sapiens*?

Homo sapiens (not always so *sapiens* but certainly improving) is a logical product of the evolutionary process. Once life began, the first very primitive molecular assemblies were the result of variations of the genetic code and therefore competed for the raw materials available for life to use. This competition led to evolution. Such evolution can be easily demonstrated today in a short time. Bacteria which are susceptible to an antibiotic will evolve quite rapidly in the presence of near lethal doses of the antibiotic. They produce a DNA plasmid which codes for an enzyme which destroys the antibiotic by a chemical reaction. Thus, they evolve their own defense system against an externally applied stress. Multicellular life is an association of the original monocellular organisms in which all the cells work together and eventually have the same genetic code. The introduction of sexuality, seen even in bacteria, permits the exchange of DNA and facilitates variations in organisms and thus helps evolution.

The change from an anerobic to an aerobic atmosphere, which took place when the blue-green algae appeared, greatly stimulated evolution. It was possible to obtain more energy from a given chemical reaction than before.

At the end of the road appeared man, a strange creature, because it had lost its fur without any apparent reason. If you wish to say that God had presided over this extraordinary story of evolution, I have no objection.

5 How should science—and the scientist—approach origin questions, specifically the origin of the universe and the origin of life?

As I have argued, the universe does not have an origin. It is a varying phenomenon. Perhaps it expands and then contracts again in an infinite number of cycles.

The origin of life, and eventually of man, does not seem to be such a problem. We have made much progress in understanding how life began and the concepts of evolution by selection can be observed in bacteria in a few days; in domestic and farm animals, as well as in plants, in a few generations.

6 Many prominent scientists—including Darwin, Einstein, and Planck—have considered the concept of God very seriously. What are your thoughts on the concept of God and on the existence of God?

As I have already stated, God is Truth. But does God really have anything to do with man? Certainly I cannot believe that God accepts only one religion, or one sect, as the only group authorized to speak for man. I would believe that God accepts all, even those who pretend not to believe. Morality and religion interact and much beneficial human behavior results from this interaction. However, one can be moral and be an atheist. Communism teaches this, and indeed real Communists consider themselves to be very moral. So religion is finally about the relationship of the individual and God. Can one speak to God? Prayers to God to advance one's personal welfare, at the expense of the less righteous, are surely not welcome. Prayers to God to let one discover truth might be acceptable. Certainly, it is remarkable how we have been able to understand so much in our environment. God permits man to make observations and experiments which can be interpreted by logical thinking. The theory that results is perpetually tested by further experiments. As a modest chemist, I cannot pretend that I have changed the world in the way that Einstein did. However, there have frequently been occasions when I was able to imagine truth that seems to have been hidden from others. Was I an instrument, of modest dimensions, chosen by God? I, of course, do not know. Also, I have no idea of life after death. Man has always wanted to believe that there is another form of life, a place where the soul would find a welcome. There is room for this hope, but it surely will not be in a form that we can, in our human frailty, imagine or understand.

4
The Mechanism of the World and the Why of It
· · · · · · · · · ·
Professor Steven L. Bernasek

- Born 14 December 1949

- Ph.D. in chemistry, University of California, 1975; received the Exxon Award in Solid State Chemistry, 1981

- Currently Professor of Chemistry, Princeton University

- Works include 79 scientific papers in refereed journals

- Professor Bernasek on:

  the origin of the universe: "The 'Big Bang' cosmology seems to serve as well as any other as far as *mechanism* for the origin of the universe".

  the origin of life: "The mechanism of the origin of life is well-supported and well-understood. The 'why' of it is again less clear".

  the origin of *Homo sapiens:* ". . . can't be distinguished *mechanistically* from the rest of the evolution process. . . . Individually, and in relation to a God, we must believe we are somehow unique".

  God: "His existence is apparent to me in everything around me, especially in my work as a scientist".

· · · · · · · · · ·

1 What do you think should be the relationship between religion and science?

My first thoughts on this question were that religion and science are somehow separate entities, but that does not fit my present thinking exactly. Perhaps my present feelings on this question assign the "mechanism" of the world to science and the "why" of it to religion. This is somewhat simplistic, but it serves. Mechanistic questions can certainly be addressed by the methods of science. In the realm of science we can propose mechanisms for how things got to be the way they are, and in most cases these propositions can be tested. There also seems to be a transience to the answers of science. Today's valid mechanism is tomorrow's childish attempt at description. I don't think religion concerns itself with these questions (nor should it). Rather, the religious tradition provides "answers" in the "why" arena based only on faith. These "answers" seem more permanent, and apparently serve well from generation to generation. The two (science and religion) are certainly connected, but they appear to operate in different realms.

2 What is your view on the origin of the universe: both on a scientific level and—if you see the need—on a metaphysical level?

The "Big Bang" cosmology seems to serve as well as any other as far as *mechanism* for the origin of the universe. I am not an expert in this area, so I can only base my answer here on what I have read in popular accounts. On a metaphysical level, the question is much more open-ended. I believe that there was some presence or being existing prior to the moment of the "Big Bang". It's hard to assign any description to this being. Perhaps the closest is in John 1:1, "In the beginning was the Word . . .".

3 What is your view on the origin of life: both on a scientific level and—if you see the need—on a metaphysical level?

Again it seems that the scientific (mechanistic) question concerning the origin of life is an easy one. The weight of scientific evidence suggests that life evolved using a carbon-based chemistry, at least on this planet. It is not clear whether life has originated in other places or other ways, although it certainly seems likely. So the mechanism of the origin of life is well-supported and well-understood. The "why" of it is again less clear. If it is *purely* the result of finite probabilities of individual chemical events and enormously long times, then there is not much else to consider. My own experience inclines me away from this conclusion, however. It is hard to accept that the mechanism is *all* there is when confronted with the enormous beauty and complexity of life on our planet. There is a lot we do not understand and perhaps never will, but that seems to be where "understanding" based on faith becomes important.

4 What is your view on the origin of *Homo sapiens*?

The origin of *Homo sapiens* can't be distinguished *mechanistically* from the rest of the evolution process. Metaphysically we certainly would like to think we are somehow unique. In the grand scheme of things it is not clear if we are or not. Individually, and in relation to a God, we must believe we are somehow unique.

This is a tough question, and I am not particularly satisfied with my own feelings on it.

5 How should science—and the scientist—approach origin questions, specifically the origin of the universe and the origin of life?

Science must continue to investigate questions of origin within the realm of science. That is to say, science must concern itself with questions of mechanism, questions of how things got this way. This certainly does not preclude the individual scientists from having religious beliefs or asking questions about the "why" of the world in the realm of religion. The two are not mutually exclusive.

6 Many prominent scientists—including Darwin, Einstein, and Planck—have considered the concept of God very seriously. What are your thoughts on the concept of God and on the existence of God?

I believe in the existence of God. His existence is apparent to me in everything around me, especially in my work as a scientist. On the other hand, I cannot prove the existence of God the way I might prove or disprove a hypothesis. My "picture" or concept is certainly colored by my Catholic upbringing and the fact that I live in a Western culture with certain historic traditions. My God is not just a kindly old man or a wrath-filled judge, though. The concept is more of a presence or sentience of what surrounds me, and a feeling of oneness with other parts of the universe.

5

The Origin of Life Seems Lost in the Details of Prebiotic Chemistry

· · · · · · · · · ·

Professor Hans J. Bremermann

- Born 14 September 1926

- Ph.D., University of Muenster, 1951

- Currently Professor of Biophysics and Mathematics, University of California, Berkeley

- Areas of specialization and accomplishments: research in quantum field theory, biophysics, several complex variables, fundamental limit of data processing; works include *Distributions, Complex Variables and Fourier Transforms*, 1965

- Professor Bremermann on:

  the origin of life: "The miracle of Newton, Einstein, and Hawking, who showed that cosmology can be derived from a few mathematical equations, so far has not repeated itself in biology. The origin of life seems lost in the details of prebiotic chemistry".

  God: "Can we equate awareness with a 'universal spirit' and the 'universal spirit' with God?"

· · · · · · · · · ·

In January 1989 I attended a two-week winter seminar "Memory, Molecules and Information", organized by Manfred Eigen. (He has long been a champion of a mathematical theory of the origin of life.) As in so many other cases, for the time being, the findings of experimental molecular biology seem to have carried the day. The miracle of Newton, Einstein, and Hawking, who showed that cosmology can be derived from a few mathematical equations, so far has not repeated itself in biology. The origin of life seems lost in the details of prebiotic chemistry. Once there are autocatalytic systems with mutability and high enough fidelity, an evolutionary path towards an "RNA world" and on to cells, metazoas, mammals, and man seems open, but we cannot reconstruct it from theory.

As to your questions about God: By and large science and religion don't mix. When theologians and church authorities issue dogmas on scientific phenomena great harm can be done to both theology and science. Conversely, good scientists often make poor theologians.

There are, however, two areas, where there may be inevitable overlaps. The observer problem in quantum mechanics, brought about by an act of observation, is apparently not relativistically invariant. When the observer is included as an interacting physical system in a relativistic quantum field theory, new problems arise: "retarded" and "advanced solutions" of the field equations exist, and throwing away the advanced solutions in the time-ordered interaction terms seems somewhat arbitrary, and in any case infinities arise which must then be removed by mathematical tricks. These tricks were first developed in the late forties, but something seems to be still amiss forty years later.

In neurobiology the human brain, like the brains from other species, is composed of individual cells and their tangled interconnections. Many neuro-computational processes go on all the time that do not rise to the level of awareness. The phenomenon of having an agent in a large parallel processor that monitors the output from subsystems and in particular the processed sensory information that describes the state of the outside world, is not only surprising, but is even necessary. That this agent should coordinate responses to the outside world likewise seems necessary under the mathematical and physical constraints that govern computation. How this thing, however, turns into the vivid, personal experience of awareness, into the sensation of color and taste and touch and smell, the experience of pleasure and pain, euphoria and depression—is a big mystery. Science can do without it, but ethics cannot ignore it.

Can we equate awareness with a "universal spirit" and the "universal spirit" with God? Here the major religions differ. In the Western religious traditions God is quite separate from the awareness phenomenon, but some Eastern traditions, if I understand them correctly, do not even have the concept of a separate deity.

I am not a theologian. Let me just say that I am profoundly appreciative of living in a society where I am allowed to live according to my conscience and to my own beliefs, without being coerced by an imposed ideology or creed. Those who love power over people will always be tempted by abusing the latter to impose their dominance over others, under the excuse that they are absolutes.

6
The Entropic versus the Anthropic Principle
. . . .
Professor Friedrich Cramer

- Born 20 September 1923

- Ph.D., University of Heidelberg; received the Kopernikus Medal, 1987

- Currently Director, Max Planck Institute for Experimental Medicine, Göttingen

- Works include *Einschlussverbindungen*, 1953; *Paperchromatography*, 1955; *Fortschritt durch Verzicht*, 1975; *Chaos und Ordnung-uber die Struktur des Lebendigen*, 1988; about 300 scientific papers in major journals on biochemistry of nucleic acids, protein-biosynthesis, and cellular interaction

- Professor Cramer on:

   the origin of the universe and of life: "Modern science, now approaching such important problems as life, the brain, evolution of the universe, and so on, has to do with systems far away from equilibrium in which irreversible thermodynamics must be applied. In these systems phenomena of self-organization are observed".

   God: "Could God exist in the concepts of science? With my new, broader concept of matter which has been briefly sketched here, I think this question can be answered positively".

. . . .

The incompatibility of the Entropic and Anthropic Principles rests upon a too-narrow concept of matter, especially of living matter. The Entropic Principle is prevalent at or near equilibrium. All classical thermodynamics, the First and Second Laws of Thermodynamics, refer to situations at and near the equilibrium and therefore deal with dead matter. Modern science, now approaching such important problems as life, the brain, evolution of the universe, and so on, has to do with systems far away from equilibrium in which irreversible thermodynamics must be applied. In these systems phenomena of self-organization are observed. In my discussion on self-organization I have shown that with the term "self-organization" one touches on the metaphysical element of a scientific evolution theory. There are no physics without metaphysical basis, but it is of utmost importance to define precisely the connecting point between physics and metaphysics in order to avoid a confusion of categories. In terms of evolution, self-organization is this connecting point between theory and metatheory. The result of closer inspection reveals then that the trivial scientific concept of matter must be sacrificed. Why not? In nuclear physics it has been sacrificed for a long time. However, these things are so abstract that they have not become common knowledge until now.

Could God exist in the concepts of science? With my new, broader concept of matter which has been sketched briefly here, I think this question can be answered positively. The biblical story of creation can be neither explained nor denied in an evolutional field theory. In the biblical creation God reveals himself and thereby does not only give an explanation of the world as we scientists try to do, but also he gives a meaning to the world. The question of meaning, however, is excluded in scientific questions and explanations by the premises on which science started.

In the evolutional field theory, matter can carry ideas. That matter could be a vehicle for numina will never be shown by science, because it is not its realm. Matter could however be this vehicle. This is no longer in contradiction with the now-proposed broader scientific concept of matter: When matter is pregnant with ideas from the beginning it could also carry God.

7

How Is It Possible to Escape the Idea of Some Intelligent and Organizing Force?

• • • • • • • • • • •

Dr. R. Merle d'Aubigne

- Born 23 July 1900

- Paris Faculty of Medicine, 1927; decorated Commander of the Legion of Honor, Croix de Guerre, Bronze Star medal

- Head of the Orthopedic Department at the University of Paris since 1949

- Areas of specialization and accomplishments: traumatic, non-traumatic lesions of hip, treatment of lesions of peripheral nerves, fractures, reparative osteogenesis, bone tumors, and new means of treating them; works include *Traumatologie*, 1955; *Traumatismes anciens*, 1958

- Dr. d'Aubigne on:

   the origin of the universe: "My mind . . . is incapable of conceiving the Nothing, before or after the existence of the universe".

   the origin of life: ". . . I cannot conceive of any physical or chemical condition where proteins could spontaneously arrange themselves in an organism bound to maintain itself with a continuous combination with oxygen and to reproduce itself".

   the origin of *Homo sapiens:* "Many facts support today the neo-Darwinian doctrine of evolution". But, "How is it possible to escape the idea of some intelligent and organizing force? This problem is likely to remain a mystery".

   God: ". . . an entity supporting the Good and the Beautiful in the emotion ineffable' their manifestations produce. . ."

• • • • • • • • • • •

1 What do you think should be the relationship between religion and science?

   NONE—except mutual respect. Science based on rationality can only *study* religions. Religion based on feeling must *accept* proved scientific facts.

2 What is your view on the origin of the universe: both on a scientific level and—if you see the need—on a metaphysical level?

No view. My mind, being a limited part of the universe, is incapable of conceiving the Nothing, before or after the existence of the universe.

3 What is your view on the origin of life: both on a scientific level and—if you see the need—on a metaphysical level?

The origin of life is still a mystery. As long as it has not been demonstrated by experimental realization, I cannot conceive of any physical or chemical condition where proteins could spontaneously arrange themselves in an organism bound to maintain itself with a continuous combination with oxygen and to reproduce itself.

4 What is your view on the origin of *Homo sapiens*?

Many facts support today the neo-Darwinian doctrine of evolution: if this theory is accepted, production of *Homo sapiens* is coherent with the appearance of mammals after progressively complex varieties of animals.

Personally, I cannot be satisfied by the idea that fortuitous mutation selected by modifications in conditions for life can explain the complex and rational organization of the brain, but also of lungs, heart, kidneys, and even joints and muscles. How is it possible to escape the idea of some intelligent and organizing force? This problem is likely to remain a mystery.

5 How should science—and the scientist—approach origin questions, specifically the origin of the universe and the origin of life?

Science has no choice: it has to study the *facts* without any preconceived view, or belief, as far and as deep its means of investigation allow it. Its field includes, of course, the origin of the universe and of life, but also the religious phenomenon in human beings.

Science may be destructive for religion because a scientific mind cannot accept many parts of religion, namely individual survival after death, with punishment or reward. But while, at the beginning of the century, some scientists thought science could substitute itself for religion, it is accepted today that scientific knowledge is limited by its very nature: take analysis of our perception by our brain cells, for instance.

A number of phenomena escape rational scientific analysis, especially emotional ones: perceptions of beauty, love, religious mysticism.

6 Many prominent scientists—including Darwin, Einstein, and Planck—have considered the concept of God very seriously. What are your thoughts on the concept of God and on the existence of God?

I suppose it is this distinction between the limited rational knowledge and the emotional perception, that has allowed many scientists (Einstein and Planck) to consider seriously the concept of God.

Very sensibly, the questionnaire solicits here our *thoughts*. It is a personal approach. If I am asked: Do you believe in God? I answer "yes". I cannot deprive myself of the feeling of an entity supporting the Good and the Beautiful, in the "emotion ineffable" their manifestations produce on me.

Whether this is a remnant of the very strong religious education I have been granted, I do not know. But it is quite strong.

8

A Divine Design: Some Questions on Origins

An Interview with Sir John Eccles (1982)

- Born 27 January 1903

- Ph.D. in natural sciences, Oxford University, 1929; Nobel Prize for Physiology/Medicine (shared with Alan Hodgkin and Andrew Huxley), 1963; received the Nobel Prize with Hodgkin and Huxley "for their discoveries concerning the ionic mechanisms involved in excitation and inhibition in the peripheral and central portions of the nerve cell membrane"

- Currently residing in Switzerland, with periodic lecture tours around the world

- Works include *Neurophysiological Basis of Mind*, 1953; *The Self and Its Brain* (with Sir Karl Popper), 1977; *The Human Psyche*, 1980

- Sir John Eccles on:

  the origin of the universe: ". . . if you look at the whole evolutionary process from the Big Bang onwards—the evolution of the cosmos and the evolution of biological life—I have a feeling that it all seems to make sense. It was as if there was a purpose in it all . . ."

  the origin of life: "The total time of fifteen billion years is really the minimum we could have gotten for making all the elements, putting them into the dust of the cosmos, and getting it all swept up eventually in our solar system which does have all the essential elements for life".

  the origin of *Homo sapiens:* ". . . the conscious self is not in the Darwinian evolutionary process at all. I think it is a divine creation".

  God: "If I consider reality as I experience it, the primary experience I have is of my own existence as a unique self-conscious being which I believe is God-created".

1 What is your conception of the Anthropic Principle, of the evidence for this principle and of its implications?

If I consider reality as I experience it, the primary experience I have is of my own existence as a unique self-conscious being which I believe is God-created. Now, if you look a the whole evolutionary process from the Big Bang onwards—the evolution of the cosmos and the evolution of biological life—I have a feeling that it all seems to make sense. It was as if there was a purpose in it all with, as Teilhard de Chardin would say, some kind of purposive goal in the whole creative process. And this leads to the incredible creation of each of us as human selves along with the whole biological side of us. The other side is the world of experience: the world of knowledge that comes via sense organs, knowledge of the world itself, of culture, art, and literature. Everything comes via sense organs to each of us and this, of course, gives rise to our own creative thoughts which come out in language and expressions of very different kinds. This is where I find myself. So how does that relate to the Big Bang? Well, there seems to be some purpose, some deeper meaning to it all. There must be a divine plan—the Anthropic Principle. This divine plan came through this whole immense cosmos.

If the cosmos had not been so immense, if we had a Big Bang and the total mass of the Big Bang was just the weight of our own galaxy, you would think this is an enormous mass for the Big Bang, the mass of our galaxy—then it all would have gone out and been back again within one year. There would have been no time for anything. The whole creation of time, and we have to have a long, long time for the creation of the elements in the supernovas and so on—all of this had to go on for thousands of millions of years. The total time of 15 billion years is really the minimum we could have gotten for making all the elements, putting them into the dust of the cosmos, and getting it all swept up eventually in our solar system which does have all the essential elements for life. You cannot make life out of hydrogen and helium, and that was the original stuff. You have to have the time for the creation of all the extraordinary elements that are necessary for living existence, and so you will have to have, shall we say, something like 10,000 million years from the Big Bang until the dust was swept up to create our solar system which only came 5000 million years ago and 4600 million years ago the earth was contracting down with not only the richness of metal, of elements and so on, but also water. Water is the most wonderful substance, and the entire earth has got it in superabundance compared to any of the other planets in the solar system.

So the creation of Planet Earth is itself a wonderful arrangement for life. The whole cosmos can be thought of as being immense in order to give time for the creation of Planet Earth and to give time for the evolution of life and eventually the creation of us in the evolutionary process. So I look upon the whole cosmic design as not being made in sheer immensity for no purpose. The sheer immensity is there in fact to get the time for the creation of Planet Earth with its immense richness of elements and then the time for it to cool down and then the time for the evolutionary process of life which took something like 3500 million years. That's the way you have to think of the time. And so this great cosmos of ours may look very extravagant in the way of material investment of mass but in order to get the long duration of the expansion you have to have this immense momentum going out in expansion, and all this is applied to time for the solar system to exist and for earth to exist and the planets too and the earth to go right through evolution and finally create us. So that is the Anthropic Principle as I see it.

2 It would seem that the Anthropic Principle, as you have outlined it, indicates that the emergence of life and of *Homo sapiens* is an immensely improbable event.

That is so. It is in fact a design, a divine design. The whole thing was wonderfully organized and planned to give the immensity, to give the size, to give the opportunity, for the Darwinist evolutionary processes that gave rise to us. So that's the Anthropic Principle. The whole process gives rise to the existence of mankind.

3 How would you respond to a naturalist interpretation of this process in terms of random events?

If you do not believe in purpose and design, then you can argue that this is just chance and necessity. But it is silly to be caught with chance and necessity for your existence. The naturalists want, on the other hand, to be leaders of thought, to be the great prophets of the age and, yet, at the same time they want to get themselves out of the process. They need a little more humility. They need the humility to think that we are all in this together—

all life and, of course, all human beings, and that they are part of the great creation plan.

4 Do you think also that what is required is a fundamental insight into the purpose and intelligibility manifested in the universe and that the naturalist lacks this insight?

I think they lack the insight because they are too proud. Monod was an example. He put himself up as a great prophet. The last chapter of his book *Chance and Necessity* in its English translation was called "The Kingdom or the Darkness", and this is a prophetic position. The Kingdom was if you went with Jacques Monod, and the Darkness was if you did not. I was in the darkness. I was an animist in his view. But this is just pride. You have to realize that we are trying to see from very inadequate positions and understanding. We must assume that we have, shall we say, supernatural powers and gifts of understanding and imagination and intelligence. We are doing the best we can to understand our situation, and when we come to that we have to be very humble in realizing that there can be much, much of greater significance that we are failing to understand, that we may gradually work through. So I want meanwhile to be very humble about our position as scientists in the natural world. Scientists have to be humble. We have not said the last word. It is the best story we have got but it has to be amended all the time. It should be regarded not as a doctrine but as a scientific hypothesis. We have to look at it all the time to see its weak points and point them out and not try to cover up the weak points. One of its weak points is that it does not have any recognizable way in which conscious life could have emerged, in which living organisms could become conscious in the evolutionary process and how in the end they could become self-conscious as we are.

5 Would you say that the main argument for theism would be the Anthropic Principle?

I think it is an important insight to have. I think it is possible to have a theistic belief based simply on our own existence. We recognize, like Descartes, that the only certainty we have is that we exist as unique self-conscious beings, each unique, never to be repeated. This I regard as outside the evolutionary process. The evolutionary process gives rise to my body and brain but, dualistically speaking, that is one side of the transaction. The other side is my conscious being itself in association with the brain for this period when I have a brain on this earth. So that brain and body are in the evolutionary process but yet not fully explained in this way. But the conscious self is not in the Darwinian evolutionary process at all. I think it is a divine creation.

6 So you see the existence of the conscious self as definite evidence for the existence of a divine Creator?

Yes, I do. I have said that several times in my books. And this is a creation, a loving creation. You have to think of it as not just by a Creator who tosses off souls one after the other. This is a loving Creator giving us all these wonderful gifts.

7 Do you see any special relevance for the theist in the Big Bang theory?

Yes, in a way, but it is the transcendent God that does it all. But when it comes to the human situation, it is the immanent God that is so important although the transcendent God is there too. We have to all the time think of them as two aspects of the same Being.

8 How can a dualist respond most effectively to a materialist who says that the brain is all there is and that we have no spiritual soul?

I would say that, when he is saying that, the materialist is making the statement that those statements are simply the brain generating lots of complicated impulses in the neurosystem and the sounds that make these words. This is entirely a material creation. You can ask, did you think it out first in thoughts and did your thoughts have anything to do with what you say? He will reply, No, what I say is simply what the brain does and I listen to it. In other words, somebody says that I don't know what I think until I listen to what I say! I think if people believe that the brain is the creator of all of their linguistic expressions and that they are merely passive recipients of the creations of their brain in language, then I don't argue with them. I don't argue with robots. I only argue with people who have a conscious, critical ability to think and to judge and to create and to recognize logic and to be defeated in argument. These are the qualities, I would say, of a human conscious person. Then you can have a dialogue, and the dialogue would be meaningful. But dialogue is not meaningful with a robot.

9

Religion and Science Neither Exclude nor Prove One Another
· · · · · · · · · ·
Professor Manfred Eigen

- Born 9 May 1927

- Ph.D. in natural sciences, University of Göttingen, 1951; Nobel Prize for Chemistry (shared with Ronald Norrish and George Porter), 1967; received the Nobel Prize with Norrish and Porter "for their studies of extremely fast chemical reactions, effected by disturbing the equilibrium by means of very short pulses of energy"

- Currently, Director of the Max Planck Institute for Biophysical Chemistry, Göttingen

- Works include *The Hypercycle: a Principle of Natural Self-Organization* (with Peter Schuster), 1979; *Laws of the Game: How the Principles of Nature Govern Chance* (with Ruthild Winkler), 1981

· · · · · · · · · ·

Your questions could be answered either with one sentence or with a whole book. The one sentence is: I think religion and science neither exclude nor prove one another. The book which I wrote two years ago answers your questions 3, 4, and 5. The title is *Stufen zum Leben*, (Munich: Piper, 1987; English translation forthcoming).

10

The Creative Process May Well Be What We Observe, Deduce, and Call Evolution

• • • • • • • • • •

Professor Thomas C. Emmel

• Born 8 May 1941

• Ph.D. in population biology, Stanford University, 1967

• Currently Professor of Zoology and Director of the Division of Lepidoptera Research, University of Florida, Gainesville

• Areas of specialization: population biology of tropical and Nearctic organisms; ecological genetics of natural populations, especially satyrid and nymphalid butterflies and land snails; territorial behavior

• Professor Emmel on:

the origin of the universe: ". . . the current Big Bang theory [is] the best explanation yet . . ."

the origin of life and of *Homo sapiens*: "To me, a timeless diety is not required by our preconceptions to create everything at once or in seven days, and then go away and disappear without further involvement in his total creation. The creative process itself may very well be what we observe, deduce, and call 'evolution'".

God: "To me, the concept of God is a logical outcome of the study of the immense universe that lies around us".

• • • • • • • • • •

1 What do you think should be the relationship between religion and science?

The ideal relationship between religion and science ought to include a mutual respect, for both are ultimately based in an intellectual curiosity about the natural world around us. Since the beginnings of his time on earth, man in the form of at least two species, *Homo sapiens* and *Homo erectus*, has wondered about his role in the world, where he came from, and other ultimate questions relating to what Margenau has called the "Ultimate Mystery that drives the religious impulse".

To me, religion addresses the spiritual issues of life—those that cannot be quantified, examined under a microscope, or approved via the scientific method, for they are ultimately involved in a sense of wonderment about questions which cannot be answered by science. The scientific enterprise, on the other hand, is stimulated by the same quest for knowledge and understanding but offers the practitioner the possibility of reaching his or her goal via the scientific method—seeing a problem, formulating a hypothesis, and being able to test that hypothesis until it is well-established as theory or even fact.

To me, then, religion and science form a linked dichotomy, in which the ultimate unifying factor is God at the base of the two branches. A tuning fork, with initially diverging branches that end up in parallel, is perhaps the best available illustration. This model, of course, does not fully express the relationship, but it does show why one who is caught up in studying the physical and biological mysteries of the world can equally appreciate the ultimate mystery responsible for its creation, its operation, and the spiritual realm with which science has no real contact. It also explains why many scientists can go down the one branch and ignore or reject the possibility of the existence of the religious branch of this dichotomous tree.

So ultimately, the relationship between religion and science should be a complementary one, in which the pursuit of one is not exclusive to the other and in which indeed the pursuit of *both* can be synergistically more productive than concentrating on either alone. Only a person supremely self-confident and egocentric could conclude that one or the other of these branches was an end in itself, without being willing to consider the merits of the other parallel track.

2 What is your view on the origin of the universe: both on a scientific level and—if you see the need—on a metaphysical level?

My view on the origin of the universe accepts the current Big Bang theory as the best explanation yet for the physical order and material construction of the universe as we see it today. Thus I believe the universe had a beginning at a moment in time some 16 billion years ago, that it started at a fixed point in space and exploded outwards, creating the galaxies and other space objects as we see them today. I see no incompatibility at all with the metaphysical inclusion of a Supreme Being starting the whole creation of the universe with such a method and at such a moment and point.

3 What is your view on the origin of life: both on a scientific level and—if you see the need—on a metaphysical level?

I agree that the scientific evidence is conclusive that the earth was formed approximately 4.6 billion years ago and that life originated some time between about 3.9 billion and 3.6 billion years ago via a model not unlike that of the Oparin-Haldane hypothesis. On the primitive earth, the simple chemicals in the reducing atmosphere created by the degassing of the earth's interior, and under the influence of natural energy sources such as lightning and ultraviolet radiation, gave rise to a series of reactive precursors of more complicated organic molecules, creating amino acids, nucleic acids, and simple sugars which could be combined in new forms to form more complex molecules. This biochemical evolution proceeded at a unique time in the earth's history, when small amounts could accumulate via concentration mechanisms such as tidal pools and not decompose, due to the absence of decay bacteria and other life.

The more advanced steps involved in the creation of the first cell, particularly the evolution of metabolism and the evolution of genetic control of protein synthesis, are steps which are very difficult to conceive. But in all of these and the earlier postulated steps, I see nothing that is incompatible with the concept of a Supreme Being ultimately being responsible for the entire act of the creation of life in this manner.

To me, *evolution* is merely a descriptive term which is applied by us to a process. This process, which we deduce has occurred in the past based on the models of present minor change that we can observe, could easily be replaced by the word "creation". In other words, creation of life by a Supreme Being could take place in exactly such a manner as in the hypotheses advanced by molecular evolutionists. After all, if we ascribe to a Supreme Being the traits of being omnipotent, omniscient, and timeless, why should man place arbitrary limitations on what that Supreme Being could accomplish in his own creation of the universe, and of our particular minor solar system and obscure planet tucked away in one corner of one of millions of galaxies? If that Supreme Being operates in his creation by his own natural laws, why *not* have a "natural" unfolding evolution of life, from its earliest precursors to today's complicated plants and animals?

To me, a timeless deity is not required by our preconceptions to create everything at once or in seven days, and then go away and disappear without further involvement in his total creation. The creative process itself may very well be what we observe, deduce, and call "evolution".

4 What is your view on the origin of *Homo sapiens*?

I agree that the origin of our most recent species of man, *Homo sapiens*, came about in the same way as other life on earth. That is, the physical body of *H. sapiens* evolved in response to a long series of selective pressures exerted on ancestral forms, including the Australopithecines of the genus *Australopithecus*, *Homo habilis*, and *Homo erectus*. The molecular genetics specialists have now shown rather convincingly and firmly that all the modern existing groups and branches of the *H. sapiens* species arose from one woman, the so-called "mitochondrial Eve", around 247,000 years ago. This single woman lived in Africa. This was neither an unexpected finding from the molecular evolutionists' point of view, nor a surprise to any natural philosopher who stopped to logically think how humans are more closely related the further we go back in time. Perhaps the only surprise was that the event of a single ancestor for all living men and women was so relatively recent in time.

It is clear from the fossil record that *H. sapiens* in a physical form evolved by about 500,000 years ago in Africa and then moved out to replace the earlier *H. erectus* populations which by then had invaded all of the Eurasian continent. Apparently, then, the most recently successful genotype of *H. sapiens* spread out from Africa starting around a quarter of a million years ago or less. Today, all of us are descended from the offspring of the single couple.

Thus it is not unreasonable to apply the biblical model of Adam and Eve to this first couple in Africa. I would point out that the Book of Genesis does make a special point of the fact that there were men on earth prior to the first man, Adam, which God "breathed the breath of life into". To me, the distinctive thing about modern man is this "soul", or capacity for a spiritual relationship with the Supreme Being, God. The Bible puts this in a poetic way (breathing the breath of life into this human body called Adam), but I see nothing incompatible with this being a poetic summary of what I have just described as the genetic origin of man in Africa. In other words, just as there was a genetic mother and father of the human species, as we see in the DNA of all humans who are alive today, there was a spiritual Adam and Eve—the first couple God revealed himself to and who realized the existence of God and subsequently decided to worship consciously his existence.

The principal incompatibility within the model that I have sketched above is simply that it seems possible that our ancestral species, *H. erectus*, had an awareness of spiritual issues prior to *H. sapiens*. Certain artifacts and deliberate careful burial of the dead by that species indicate a belief in a life after death. However, the dating is uncertain (these examples could date as recently as 250,000 years ago for *H. erectus*), and these spiritual figures or other artwork and deliberate burial practices may relate to later *H. erectus* populations who could have picked up these concepts from sympatric *H. sapiens* groups. Perhaps we will never know the truth of the matter with absolute certainty, but the above model is an interesting way to show the possible complete compatibility of the origin-of-man question with both the scientist and the theologian.

5 How should science—and the scientist—approach origin questions, specifically the origin of the universe and the origin of life?

I think the origin of the universe and the origin of life are two of the ultimate questions we can ask, either as theologians or as scientists. The scientist should approach the question as reverently as the theologian, that is, apply the best of his intelligence and perspective to the problem and not be inhibited by the quest or where it may take him. To study these two questions, however, one must be a well-read *generalist,* and that is rare indeed among today's scientists. It is perhaps humanly impossible to master all branches of science which must be applied to these two questions, from physics to astronomy to geology to chemistry to biochemistry to ecology to evolutionary theory to anthropology, and so on.

Once, several centuries ago or even in the beginning of this century, a scientist could attempt to read all the literature published each year in science and claim he or she had some general base of knowledge broad enough to adequately tackle these questions alone. Now, it is probably best that a team of scientists or at least a *group* of frequently interacting colleagues debate the questions and consider the evidence in developing new theories and lines of research. To me, keeping open the religious perspective during such a quest broadens the search and enables the scientist to think with a clearer mind and a broader mind about the proper questions to ask and the answers to seek.

6 Many prominent scientists—including Darwin, Einstein, and Planck—have considered the concept of God very seriously. What are your thoughts on the concept of God and on the existence of God?

To me, the concept of God is a logical outcome of the study of the immense universe that lies around us. My readings in science and my professional pursuit of science have simply confirmed further to me that there are ultimate questions that we as scientists cannot answer, but yet these questions exist and it is silly at best, and unscientific at worst, to ignore them.

The fact that there has been a drive in every society of which we have record towards understanding the ultimate mystery that drives this religious impulse is something that we as scientists have to take very seriously and not ignore. I feel that many scientists reach a point during their graduate student days or perhaps a little later in which they feel it is *unfashionable* to consider metaphysical views, and so they bury their heads in the sand for the rest of their lives, not making any effort to see a perspective broader than their own immediate field. Great pioneers such as Isaac Newton, Charles Darwin, Albert Einstein, and Max Planck have considered the concept of God and his existence very seriously, and I feel it is rewarding to pursue the same line.

To me, God exists as the Supreme Being who started this creation that we call the universe, and he was responsible in a very real sense for the development of life on earth which ultimately led to man. The evidence is all-too-pervasive for me to think otherwise. I do not pretend to believe that I will ever know the ultimate mysteries of this Supreme Being, but then it would be hard to imagine a truly unlimited Supreme Being who *could* be understood by an extremely minor part of his creation.

11

At Some Stage in Evolution, God Created the Human Soul

· · · · · · · · · ·

Professor P. C. C. Garnham

- Born 15 January 1901

- M.D., University of London, 1928; D.Sc., University of London, 1952; received the Darling Medal and Prize, 1951; Bernhardt Noct Medal, 1957; Gaspar Vianna Medal and Decoration, 1962; Manson Medal, 1964

- Currently Emeritus Professor of Medical Protozoology, University of London

- Works include *Immunity of Protozoa* (with Roitt Pierce), 1965; *Malaria Parasites*, 1966; *Catalogue of Garnham Collection of Malaria Parasites*, 1981; about 400 scientific papers in various scientific journals

- Areas of specialization and accomplishments: discovered third cycle malaria parasites in liver of monkeys and man, ultrastructures in malaria parasites and other parasitic protozoa

- Professor Garnham on:

  the origin of the universe: "God originated the universe or universes".

  the origin of life: "The answer . . . is surely God".

  the origin of *Homo sapiens*: "At some stage in evolution, when proto-humans were sufficiently advanced, God created the human soul".

  God: ". . . God must exist."

· · · · · · · · · ·

1 What do you think should be the relationship between religion and science?

A relationship should be encouraged between religious and scientific thinking with agreement between them, the former basing their outlook on *faith* and the latter on *scientific observations*, with the proviso that both methods are valid.

2 What is your view on the origin of the universe: both on a scientific level and—if you see the need—on a metaphysical level?

> God originated the universe or universes. The direct or literal interpretation of the biblical accounts in Genesis, both in regard to time or details, are unacceptable to the scientific and to much of the religious mind. Such comments apply to the following questions. Theories of the origin of the universe change, but many people accept provisionally Stephen Hawking on the genesis of the universe, from the Big Bang to black holes— though some of his theological assumptions do not seem to many of us to be valid—and it should be remembered that he only deals with *one* universe, though accepting the existence of others.

3 What is your view on the origin of life: both on a scientific level and—if you see the need—on a metaphysical level?

> The answer to your question is surely "God".

4 What is your view on the origin of *Homo sapiens*?

5 How should science—and the scientist—approach origin questions, specifically the origin of the universe and the origin of life?

> The origin of *Homo sapiens*: at some stage in evolution, when proto-humans were sufficiently advanced, God created the human soul. It seems reasonable to think that God instilled a similar structure in all living members of the animal and vegetable kingdoms.
>
> The serious defect in the attitude of most scientists is their refusal to consider the "paranormal and extra sensory perception". (This is outside your range of enquiries, but the reality or otherwise of these so-called phenomena should be of interest both to the scientific and religious mind—surely they have a place for consideration by the "open mind".)

6 Many prominent scientists—including Darwin, Einstein, and Planck—have considered the concept of God very seriously. What are your thoughts on the concept of God and on the existence of God?

> By faith and by appreciation of scientific necessity, God must exist.

12

A Spirit Which Has Established the Universe and Its Laws

· · · · · · · · · ·

Professor Roger J. Gautheret

- Born 29 March 1910

- Ph.D., Paris Faculty of Sciences, 1935

- Professor of Cell Biology at the Paris Faculty of Sciences until 1979 and President of the Academy of Sciences, Paris, 1979 to 1980

- Area of specialization: plant tissue

- Professor Gautheret on:

   the origin of the universe: "The Big Bang and the expansion hypothesis assign a time-limit in the past. But what was before?"

   the origin of life: ". . . life is inseparable from cell organization. . . . The codification by chance of such a system is almost totally unlikely and nevertheless it has been realized".

   the origin of *Homo sapiens*: "Organism improvement suggests the God concept".

   God: ". . . notions such as infinite space and time, matter, structure, and order which govern the universe suggest the intervention of a spirit which has established the universe and its laws".

· · · · · · · · · ·

1 What do you think should be the relationship between religion and science?

The progress of science, especially of physics, cosmology, and biology, leads us more and more to impasses of human understanding and which can be accessible only by belief. These impasses which engage the human spirit would be cleared only by an intelligence much higher than man's understanding. This leads finally to the notion of God. The connections between science and religion were bad during the sixteenth century and worst during the nineteenth. These connections would be much better since it is clear that the human understanding of every-thing is limited. Religion could at least, from a philosophical point of view, help to surpass the limits.

2 What is your view on the origin of the universe: both on a scientific level and—if you see the need—on a metaphysical level?

Presently we don't know if the universe is infinite or limited in space and time. The notion of the infinite is not accessible to human understanding. The Big Bang and the expansion hypothesis assign a time-limit in the past. But what was before? With these hypotheses the problem is not solved but evaded. As a matter of fact the Big Bang expresses man's anxiety about the infinite.

3 What is your view on the origin of life: both on a scientific level and—if you see the need—on a metaphysical level?

Paleontology assigns clearly the geological period life appeared on earth. But we don't know how this appearance happened. Our opinion can be suggested only by the present situation. Now, life is inseparable from cell organization. One cell includes numerous kinds of enzymes, and the lack of a single enzyme may prevent life. The codification by chance of such a system is almost totally unlikely and nevertheless it has been realized. This is the main mystery of biology.

4 What is your view on the origin of *Homo sapiens*?

The progress of anthropology has shown that man is not funda-mentally different from other living organisms. He is the last stage of evolution. But there is no reason to believe that the evolution is finished. We must believe that organisms will mod-ify and transform into others as in the past. Then, observing the fossils we must admit that evolution presented the characteris-tics of a progressive improvement, especially in the case of ani-mals. It is hardly probable that this evolution has proceeded by chance, for it seemed to follow a rather rigorous pattern. Organ-ism improvement suggests the God concept.

5 How should science—and the scientist—approach origin questions, specifically the origin of the universe and the origin of life?

After an extraordinary development, one can see the limits of science except in the cases accessible to experimentation. No experimentation is possible in cosmology, and when all the observations will be done with more and more powerful instruments, the subject will become philosophical. Concerning the origin of life it must not be impossible to realize the conditions which were existing when the first living organism appeared. This is not a philosophical problem. And if the recreation of these conditions led to spontaneous generation, it would be useless to invoke the intervention of God. But presently we are very far from knowing what these conditions were.

6 Many prominent scientists—including Darwin, Einstein, and Planck—have considered the concept of God very seriously. What are your thoughts on the concept of God and on the existence of God?

To sum up, I believe that notions such as infinite space and time, matter, structure, and order which govern the universe suggest the intervention of a spirit which has established the universe and its laws. Reflection on these subjects cannot avoid the notion of God.

13

I Have a Religious Attitude toward the Unknown

Professor Ragnar Granit

- Born 30 October 1900

- Medical degree, University of Helsinki, 1927; Nobel Prize for Physiology/Medicine (shared with H. Keffer Hartline and George Wald), 1967; received the Nobel Prize with Hartline and Wald "for their discoveries concerning the primary physiological and chemical visual processes in the eye"

- Director of the Nobel Institute for Neurophysiology and Professor at the Karolinska Institute until retirement in 1967

- Works include: *Receptors and Sensory Perceptions*, 1955; *The Basis of Motor Control*, 1970; *The Purposive Brain*, 1977

- Professor Granit on:

    the origin of the universe: "Believing in the Big Bang theory, what is the origin of the situation thus created to lead to it?"

    the origin of *Homo sapiens*: "I am a Darwinian, even though convinced that neo-Darwinism only contains parts of the full explanation".

    God: "Aware of the limitations of science, I have a religious attitude toward the unknown".

1 What do you think should be the relationship between religion and science?

The relationship should be benevolent on both sides.

2 What is your view on the origin of the universe: both on a scientific level and—if you see the need—on a metaphysical level?

Believing in the Big Bang theory, what is the origin of the situation thus created to lead to it?

3 What is your view on the origin of life: both on a scientific level and—if you see the need—on a metaphysical level?

I am a Darwinian, even though convinced that neo-Darwinism only contains part of the full explanation.

4 What is your view on the origin of *Homo sapiens*?

This also covers your Question 4.

5 How should science—and the scientist—approach origin questions, specifically the origin of the universe and the origin of life?

Humbly.

6 Many prominent scientists—including Darwin, Einstein, and Planck—have considered the concept of God very seriously. What are your thoughts on the concept of God and on the existence of God?

Aware of the limitations of science, I have a religious attitude toward the unknown.

14

I Consider the Existence of God as Unknowable

Professor Robert W. Holley

· · · · · · · · · ·

- Born 28 January 1922

- Ph.D. in chemistry, Cornell University, 1947; Nobel Prize for Physiology/Medicine (shared with Har Gobind Khorana and Marshall W. Nirenberg), 1968; received the Nobel Prize with Khorana and Nirenberg "for their interpretation of the genetic code and its function in protein systhesis"

- Currently Resident Fellow and American Cancer Society Professor of Molecular Biology, The Salk Institute

- Professor Holley on:

  the origin of the universe: "I have no reason to challenge the Big Bang hypothesis".

  the origin of life: "The step-by-step development of organic molecules. . . and so forth seems reasonable to me".

  the origin of *Homo sapiens:* ". . . developed from ancestors closely related to the large apes".

  God: ". . . I consider the existence of God as 'unknowable', and therefore part of one's religious view".

· · · · · · · · · ·

1 What do you think should be the relationship between religion and science?

Religion deals with the "unknowable", science with the "knowable". Conflicts arise when people think something that has been "unknowable" has become "knowable".

2 What is your view on the origin of the universe: both on a scientific level and—if you see the need—on a metaphysical level?

I have no reason to challenge the Big Bang hypothesis.

3 What is your view on the origin of life: both on a scientific level and—if you see the need—on a metaphysical level?

The step-by-step development of organic molecules, nucleotide-like molecules, polynucleotides, polypeptides, nucleic acids, proteins, enzymes, enzyme complexes, cells, combinations of cells, and so forth seems reasonable to me.

4 What is your view on the origin of *Homo sapiens*?

Homo sapiens developed from ancestors closely related to the large apes. The genetic makeup of man is almost identical to that of the large apes.

5 How should science—and the scientist—approach origin questions, specifically the origin of the universe and the origin of life?

If he thinks the answer is "knowable", he should approach the questions as a scientist.

6 Many prominent scientists—including Darwin, Einstein, and Planck—have considered the concept of God very seriously. What are your thoughts on the concept of God and on the existence of God?

I consider the existence of God as "unknowable", and therefore part of one's religious view. There is a great deal to marvel or wonder about in the universe. Whether one wants to attribute the marvelous things to the existence of God depends upon one's nature and experience. Such a belief appeals to some people and not to others. Since it is unknowable, I think it should be a very personal matter.

15

I Have No Way of Knowing Whether God Exists

* * * * * * * * * *

Professor Jerome Karle

- Born 18 June 1917

- Ph.D. in physical chemistry, University of Michigan, 1943; Nobel Prize for Chemistry (shared with Herbert A. Hauptmann), 1985; received the Nobel Prize with Hauptmann "for their outstanding achievements in the development of direct methods for the determination of crystal structures"

- Currently chief scientist of the Laboratory for the Structure of Matter at the Naval Research Laboratory

- Works include *Solution of the Phase Problem I: The Centrosymmetric Crystal* (with Herbert A. Hauptmann), 1953

- Professor Karle on:

  the origin of the universe: "I am prepared to accept [the Big Bang model]. . . ."

  the origin of life: Arose as a result of appropriate "chemical and physical circumstances". But ". . . the kind of evidence required to build up a clear understanding is very difficult to acquire".

  the origin of *Homo sapiens*: "The evidence for evolution is compelling".

  God: "If one wants to call the beauty and magnificence of nature and the highest creativity and ethical behavior of human beings as manifestations of what they mean by God, I can be comfortable with that. In that sense, the concept of God would be the composite of the highest experiences that man could perceive in his existence".

* * * * * * * * * *

1 What do you think should be the relationship between religion and science?

To the extent that many religions concern a high form of ethical behavior and the fact that science must not tolerate anything but the highest form of ethical behavior among its practitioners, there is an overlap of interest. Beyond that, religions are perfused

181

with beliefs that are not supported by evidence. One of the consequences is that our world is replete with a variety of religions and a variety of beliefs. I personally have no problem with that except when beliefs are imposed on those who do not agree or when beliefs are used to justify unethical or violent behavior. Put simply, I expect civilized people to behave in a humane fashion, no matter what they think to themselves.

In contrast to faith and beliefs, science ultimately progresses on the basis of evidence. Believing or hypothesizing for a scientist is only the first step. A belief becomes part of science when evidence accumulates to support the belief. As scientists we are tremendously curious and awestruck by the workings and beauty of nature. It is important, however, not to translate awe into metaphysical explanations of events.

Whether the human mind is capable of comprehending the nature of the universe remains a question. We certainly can understand some things. The extent to which we can discover and understand is satisfying enough to me.

2 What is your view on the origin of the universe: both on a scientific level and—if you see the need—on a metaphysical level?

I read with interest progress in the science of cosmology which is trying to tell us, on the basis of astrophysics and astronomical measurements, what the evolutionary history of the universe is. I have nothing to add to it. If the present course of observational events arose from a giant explosion ten to twenty billion years ago, as is indicated by current evidence, I am prepared to accept it, although, as Einstein said, it seems preposterous. Cosmology is a very fine example of a field of science that is in its earliest stages of development. Its progress in recent years has been considerable and there is every reason to believe that its progress will continue. As insights into the origin of the universe develop, my views will also be so affected.

3 What is your view on the origin of life: both on a scientific level and—if you see the need—on a metaphysical level?

Life arose because the chemical and physical circumstances were such that combinations of chemical entities could take on those characteristics that the biologists would define as appropriate to living organisms such as replication, deriving energy from surroundings, and so on. Some progress has been made in reproducing the primitive conditions under which the proper chemical compounds may have been synthesized. There is much work required in this area since it appears that the kind of evidence required to build up a clear understanding is very difficult to acquire.

4 What is your view on the origin of *Homo sapiens*?

In the context of what I have already said about the origin of the universe, I look to the evolutionary biologists, the paleontologists, and the archeologists to find and sift the evidence regarding the origin of humans. I happen to have a graduate degree (M.A.) in biology from Harvard University. As a student of biology I never felt the need to distinguish the evolutionary origins of humans from that of their other living predecessors. The evidence for evolution is compelling.

5 How should science—and the scientist—approach origin questions, specifically the origin of the universe and the origin of life?

Science and the scientist should approach origin questions in the same way that they approach any other question concerning the universe, namely, by hypothesizing and gathering evidence. If no evidence is available, there is no science and one is left with mythology, which I do not find interesting.

6 Many prominent scientists—including Darwin, Einstein, and Planck—have considered the concept of God very seriously. What are your thoughts on the concept of God and on the existence of God?

In order to have thoughts that interest me at all, a subject has to be sufficiently well-defined so that the possibility of gathering evidence for the subject's existence and behavior can be considered. I am aware of many commonplace descriptions of God that emphasize human features and also omniscience and omnipotence. I have no way of knowing whether such a being exists or not. If one wants to call the beauty and magnificence of nature and the highest creativity and ethical behavior of human beings as manifestations of what they mean by God, I can be comfortable with that. In that sense, the concept of God would be the composite of the highest experiences that man could perceive in his existence.

16

A Religious Impulse Guides Our Motive in Sustaining Scientific Inquiry

• • • • • • • • • •

Professor Joshua Lederberg

• Born 23 May 1925

• Ph.D. in microbiology, Yale University, 1948; Nobel Prize for Physiology/Medicine (shared with George W. Beadle and Edward L. Tatum), 1958; received the Nobel Prize "for his discoveries concerning the genetic recombination and the organization of the genetic material of bacteria"

• Currently President of the Rockefeller University

• Works include *Papers in Microbial Genetics*, 1951; *Man and His Future*, 1962.

• • • • • • • • • •

Religion and Science

The framework of reply that you offer is, I find, too narrow. Science has a long way to go in completing *its* account of the history of the universe and the origin of life and I see no point in pitting that process against one labelled as "religious". What is incontrovertible is that a religious impulse guides our motive in sustaining that scientific inquiry. If not, what else?

17
New Visions of the Cosmos
· · · · · · · · · ·
Professor Clifford N. Matthews

- Born 20 December 1921

- Ph.D. in chemistry, Yale University, 1955

- Currently Professor of Chemistry, University of Illinois, Chicago

- Areas of specialization: chemical evolution, origin of molecules in biochemistry, geochemistry, and galactochemistry, cosmochemistry and the origin of life

· · · · · · · · · ·

A true science of life must let infinity in . . .
    —*Arthur Koestler*

Science

- an endless search for unity in nature
- a continuing probe into the mystery of order
- a wandering dialogue with the unknown

is in our time achieving a new universality through the revelation that life may be an inherent property of matter arising by continuous processes of

- chemical evolution
- biological evolution
- cultural evoluton

making possible today our re-entry into nature, here on earth and in the cosmos as a whole.

From wonder into wonder, existence opens.
    —*Laotzu*

In our evolving universe

- participatory
- transcendent
- open

we come to terms daily with the ultimate mystery of existence, through our creative activities in art, science, and religion that produce the metaphors, models, and myths by which we live.

Our cosmos bears the imprint of our minds. . . . All science is cosmology.

—*Karl Popper*

18

Nobel Prize Winners Are Not More Competent about God

Professor Vladimir Prelog

• • • • • • • • • •

- Born 23 July 1906

- Ph.D. in science and technology, Institute of Technology, Prague, 1929; Nobel Prize for Chemistry (shared with John W. Cornforth), 1975; received the Nobel Prize "for his research into the stereochemistry of organic molecules and reactions"

- Professor of Organic Chemistry at the Federal Institute of Technology, Zurich, until 1976

• • • • • • • • • •

I write this note with some hesitation and repeat some remarks which I made a few years ago in a letter to a Miss Gillen in New York. Winners of the Nobel Prize are not more competent about God, religion, and life after death than other people; some of them, like myself, are agnostics. They just don't know, and therefore they are tolerant of religious people, atheists, and others.

My agnosticism goes so far that I am not even certain whether I am an agnostic. I am pleased, for instance, by Faust's answer to Gretchen's question. I often cite Max Planck: God is at the beginning of every religion and at the end of the natural sciences. I am also in agreement with the clever rabbi who said, "God is so great that he does not need to exist". I follow with interest the discussions concerning the origin of the universe, of life, and of man, especially in the molecular area. I marvel at the courage of the scientists who deal with questions of the origin and seek answers for them. I only fear that our knowledge concerning the physical, chemical, and biological (as well as the psychological and epistemological) bases does not suffice to give presently satisfying answers.

The search for these foundations (for instance, the structure of matter or the molecular evolution) which, as Isaac Rabi said, brings us closer to God, is in my opinion the noblest task of the sciences.

19

The Universe Started from an Instability in the Quantum Vacuum

. . . .

Professor Ilya Prigogine

- Born 25 January 1917

- Ph.D. in chemistry, Free University of Brussels, 1945; Nobel Prize for Chemistry, 1977; received the Nobel Prize "for his contributions to nonequilibrium thermodynamics, particularly the theory of dissipative structures"

- Currently Director of the Solvay International Institute of Physics and Chemistry, Brussels, and Director of the Ilya Prigogine Center for Statistical Mechanics and Thermodynamics, Austin, Texas

- Works include *Introduction to Thermodynamics of Irreversible Processes*, 1962; *From Being to Becoming: Time and Complexity in the Physical Sciences*, 1980; *Order Out of Chaos* (with Isabelle Stengers), 1984

- Professor Prigogine on:

  the origin of the universe: ". . . the universe started from an instability of the quantum vacuum".

  the origin of life: ". . . life is a cosmic phenomenon, which appears where the right conditions . . . are satisfied".

  God: ". . . starting with the testimony of paleolithic art, we see that man has always tried to separate truth from error and meaning from appearance. This fundamental dimension of human consciousness applies both to the scientific enquiry as well as to the human feeling of transcendence, as manifest in religions".

. . . .

1 What do you think should be the relationship between religion and science?

We are going through a period of an extraordinarily rapid evolution of science. Not only are the domains to which we may apply scientific enquiry expanding rapidly; but the image of our universe is changing. Our vision of nature is undergoing a radical change towards the multiple, the temporal, and the complex. Curiously the unexpected complexity which has been discovered in nature has not led to a slowdown of science but, on the contrary, to the emergence of new conceptual structures that now appear as essential to our understanding of the physical world—the world that includes us. I have tried to express this point of view in my book (Ilya Prigogine and Isabelle Stengers, *Order out of Chaos*, New York: Bantam, 1984). This situation inspires in us an increased feeling of astonishment in front of this nature of which we are part. It is only natural that this astonishment takes so many forms; in these circumstances, we should emphasize tolerance among different world views, of which traditional religious world views and traditional materialistic world views would be only two examples.

2 What is your view on the origin of the universe: both on a scientific level and—if you see the need—on a metaphysical level?

I do not believe that we can speak about an origin of our universe. Science can never deal with unique events, but only with classes of events. Instead of a Big Bang, corresponding to an initial singularity, I believe that the universe started from an instability of the quantum vacuum. This type of instability can arise again and may have arisen in the past over indefinite periods of time. It seems to me that in this perspective, time precedes existence, as the universe appears as an actualization of large scale irreversible processes implying an arrow of time. (More details may be found in a recent paper [I. Prigogine, J. Geheniau, E. Gunzig, & P. Nardone, Proc. Natl. Acad. Sci. USA 85, 7428–7432 (1988)]).

3 What is your view on the origin of life: both on a scientific level and—if you see the need—on a metaphysical level?

Again, I do not believe in life as a unique phenomenon. I believe life is a cosmic phenomenon, which appears where the right conditions (distance from equilibrium, non-linear processes, chemical composition, and so on) are satisfied. The main difficulty with the classical view of its origin was the high "improbability" of the complex organization which we find in living beings. With the discovery of complex structures arising in non-equilibrium non-linear systems, this objection can be discarded. In a hydrodynamic process such as Benard instability, we see complex collective phenomena which at equilibrium would be quite improbable, too. But the non-equilibrium provides the necessary environment for these self-organizing processes and permits the realization of processes whose probability at equilibrium would be nil. Non-life and life are likely to be separated through a number of instabilities, leading to bifurcations or non-equilibrium phase transitions which we may hope to unravel over the next years. A recent book gives in my opinion a balanced view on the role of chaos and bifurcations in living systems (Friedrich Cramer, *Chaos Und Ordnung,* Stuttgart: DVA, 1988).

4 What is your view on the origin of *Homo sapiens?*

I have no opinion on this subject.

5 How should science—and the scientist—approach origin questions, specifically the origin of the universe and the origin of life?

I have presented some comments on this question above; we probably never will have a final answer. Each generation has to reconsider these problems, taking into account the state of knowledge. An important element is the transition from a description involving interactions, elementary particles, molecules, and so on, to one involving the concept of information. Here, classification of dynamical systems, as introduced over the last decades, is likely to play an important role. We know now how chaos may generate information. This point is developed in a book I have written with one of my colleagues (Ilya Prigogine and Gregoire Nicolis, *Exploring Complexity,* New York: Freeman, forthcoming). As I mentioned, concepts such as self-organization, chaos generating information, have to play an essential role in such problems.

6 Many prominent scientists—including Darwin, Einstein, and Planck—have considered the concept of God very seriously. What are your thoughts on the concept of God and on the existence of God?

One of the main changes in our view about the universe is the fact that we have to give up the myth of complete knowledge. In the frame of the new theory of dynamical systems that is emerging now, we only have a finite window on the state of a system. Classical descriptions of mechanics, involving concepts such as trajectories, have not the generality which we ascribed to them. For unstable systems, the concept of trajectory becomes an unrealistic idealization. I subscribe entirely to the declaration of J. Lighthill ("The Recently Recognised Failure of Predictability in Newtonian Dynamics", in *Predictability in Science and Society,* edited by J. Mason et. al., a special issue of the Proceedings of the Royal Society, A407, 35–60 [1986]): "I . . . have to speak . . . on behalf of the broad global fraternity of practitioners of mechanics. We collectively wish to apologize for having misled the general educated public by spreading ideas about the determinism of systems satisfying Newton's laws of motion that, after 1960, were to be proved incorrect". It is quite a unique event that we have to revise ideas which have endured for three centuries. I may mention that a similar evolution in our ideas about quantum mechanics is necessary. Here also, for well-defined classes of dynamical systems, the concept of wave function becomes an unrealistic limit, as the knowledge of the wave function corresponds then to the limit of infinite information.

Herbert Simon has proposed the idea of bounded rationality, which I find quite appropriate. He writes:

> The term "rational" denotes behaviour that is appropriate to specific goals in the context of a given situation. If the characteristics of the choosing organism are ignored, and we consider only those constraints that arise from the external situation, then we may speak of substantive or objective rationality—that is, behaviour that can be adjudged objectively to be optimally adapted to the situation. On the other hand, if we take into account the limitations of knowledge and computing power of the choosing organism, then we may find it incapable of making optimal choices. If however, it uses methods of choice that are as effective as its decision-making and problem-solving means permit, we may speak of procedural or bounded rationality.

(See Herbert Simon, "Human Nature in Politics: The Dialogue of Psychology with Political Science", reprinted in Miriam Campanella, ed., *Between Rationality and Cognition: Policy-Making under Conditions of Uncertainty, Complexity and Turbulence*, Torino: Albert Meynier, 1988, pp. 11–34 [p. 13]).

This idea as proposed by Simon has emerged from "soft sciences"; however, I find it quite consonant with the recent evolution of our ideas about dynamical systems, classical or quantum. Bounded rationality does not mean irrationality; it is not a defeat of the human mind, but a new proof of our capability to deal with problems of a much greater complexity than classical science could envisage. This reconceptualization of physics may eventually lead to new ways of conceiving relations of man with nature and of man with divinity. Mircea Eliade has said that we have in human history an alternation between periods emphasizing the idea of a personal God, outside the world he has created, and periods where mankind believes in a more pantheistic nature, in which the distinction between God and nature is blurred. Whatever this may be, starting with the testimony of paleolithic art, we see that man has always tried to separate truth from error and meaning from appearance. This fundamental dimension of human consciousness applies both to the scientific enquiry as well as to the human feeling of transcendence, as manifest in religions. I believe we are approaching a time where, more than the relations of God with man, the relations of man with nature (which gave birth to man) is becoming an essential element of our perception of existence.

20
It Is Probably Impossible to Explain a Miracle with Physics and Chemistry

• • • • • • • • • •

Professor Tadeus Reichstein

• Born 20 July 1897

• Ph.D. in organic chemistry, Federal Institute of Technology, Zurich, 1922; Nobel Prize for Physiology/Medicine (shared with Philip S. Hench and Edward C. Kendall), 1950; received the Nobel Prize with Hench and Kendall "for their discoveries relating to the hormones of the adrenal cortex, their chemical structure and biological effects"

• Currently Professor Emeritus at the University of Basel

• • • • • • • • • •

The questions you raise are fundamental for humanity. Personally I very much agree with [the] concepts of Darwin, Einstein, Schrödinger, Heisenberg, Planck, and others. I do not think it makes sense to take statements of the Bible as purely rational events. The book will remain, nevertheless, a fundamental expression of the human mind, speaking in the language of the unconscious and, therefore, including the irrational side. As far as I can see, Darwin was a very religious man; the Darwinists misinterpreted his attitude. It is probably impossible to explain a miracle with physics and chemistry.

21

I Have Very Little in the Way of Belief in a Concrete God

• • • • • • • • • • •

Professor Frederick C. Robbins

- Born 25 August 1916

- M.D. in pediatrics, Harvard Medical School, 1940; Nobel Prize for Physiology/Medicine (shared with John F. Enders and Thomas H. Weller), 1954; received the Nobel Prize with Enders and Weller "for their discovery of the ability of poliomyelitis viruses to grow in cultures of various types of tissue"

- Currently University Professor Emeritus at Case Western Reserve University

- Professor Robbins on:

  the origin of the universe: ". . . we have some scientific evidence that is suggestive . . . as to how the universe was originally formed". But ". . . there is always the unanswerable question of what went before that".

  the origin of life: ". . . it probably began when certain elements fused into large molecules which eventually developed into the extraordinary system we know today".

  the origin of *Homo sapiens:* "I think the concepts of evolution, both chemical and biologic, are quite adequate to explain the origin of *Homo sapiens* . . .".

  God: "No matter how deeply we probe scientifically, I doubt if we will be able to discover the ultimate answers. . . . I have very little in the way of belief in a concrete God".

• • • • • • • • • •

1 What do you think should be the relationship between religion and science?

It seems to me that scientists can concern themselves with religion as individuals, but science cannot really deal with religion. Science is trying to explain the knowable, whereas religion is dealing with the unknowable. Of course, one can always argue over definitions. To me, science is experimental science. If one defines science as knowledge, then perhaps one might take a somewhat different view. However, religion is based primarily on faith, whereas one tries to base knowledge and science on evidence.

2 What is your view on the origin of the universe: both on a scientific level and—if you see the need—on a metaphysical level?

It seems to me that we have some scientific evidence that is suggestive, at least, as to how the universe was originally formed. Of course you know that there is always the unanswerable question of what went before that. Personally, I think it is interesting to explore the question of origin scientifically. I have no interest in pursuing it to the metaphysical level. I am quite happy to say I don't know.

3 What is your view on the origin of life: both on a scientific level and—if you see the need—on a metaphysical level?

Again my own view on the origin of life is that it probably began when certain elements fused into large molecules which eventually developed into the extraordinary system we know today. I'm not particularly persuaded by the idea expressed in Genesis.

4 What is your view on the origin of *Homo sapiens*?

I think the concepts of evolution, both chemical and biologic, are quite adequate to explain the origin of *Homo sapiens*?

5 How should science—and the scientist—approach origin questions, specifically the origin of the universe and the origin of life?

It seems to me that they should approach them scientifically and search for evidence. The experimental approach to the origin of complex molecules which might precede living things seems to me a reasonable approach. The search for such molecules elsewhere in the universe, the use of geological, astronomical, and biological data and others to help verify evolution, is a proper scientific approach. One can speculate, one can adopt a religous approach and have faith, but to my way of thinking this is not scientific.

6 Many prominent scientists—including Darwin, Einstein, and Planck—have considered the concept of God very seriously. What are your thoughts on the concept of God and on the existence of God?

Man has felt the need through the ages to believe in some higher being or beings in order to explain what happens around him and to seek out comfort. When one lives in a miserable hovel and is faced with possible starvation, the thought of heaven and a merciful God can make life bearable. No matter how deeply we probe scientifically, I doubt if we will be able to discover the ultimate answers. One can, if one wishes, adopt the solution that is God, but even this doesn't answer very much. Personally, I have very little in the way of belief in a concrete God. I accept the mystery and have adopted a rather simple-minded philosophy of living, which is that the earth is peopled with living beings, and that in order to survive there are certain behaviors which would be most favorable and it is these that I would like to follow and have my fellows follow. To some this would be religous practices. I simply don't ascribe them to God or a supreme being. I don't rule out the possibility of some God-like force, but it is beyond my capability of conceptualizing.

22

The Piling of Coincidence on Coincidence

· · · · · · · · · ·

Professor Jay Roth

- Born 10 June 1900

- Ph.D. in organic chemistry, Purdue University, 1944

- Currently Professor of Cell and Molecular Biology, Emeritus, University of Connecticut, Storrs

- Areas of specialization: nucleic acids and nucleases in relation to cell division; control mechanisms in cancer; deoxynucleotide metabolism in growth; virus tumor biochemistry

- Professor Roth on:

  the origin of the universe: "There is so much in the physical nature of the universe we inhabit, the exact balances of everything needed to support life, the piling of coincidence on coincidence, every one of which is vitally necessary for development of a stable star with a planet that can support life. These physical properties of the universe lead me to favor a Designer or Creator . . ."

  the origin of life: "The original template (be it DNA or RNA) . . . must have been very complex indeed. For this template and this template alone, it appears it is reasonable at present to suggest the possibility of a Creator".

  the origin of *Homo sapiens:* ". . . we evolved from the primates by a series of fortuitous changes which are unknown".

  God: ". . . for the circumstantial reason stated, I believe in a creator".

· · · · · · · · · ·

1 What do you think should be the relationship between religion and science?

I think the relationship should be one of mutual respect and tacit agreement not to attack the other. The value of faith to large numbers of people is apparent. Faith in a divine being or beings, faith in an ultimate purpose to life, faith in a (pleasant) afterlife can be of enormous comfort. Faith in the teachings of religion can be a strong force for good in the world.

With the exception of fundamentalists, most clergy recognize the essential nature of science and its quest for truth and understanding of the natural world. The informed clergy have no quarrel with scientific belief and knowledge and may, in fact, agree that the Bible is more symbolic than literal, but, nevertheless, important for its moral precepts.

Adherents of both science and religion should recognize that while both are basically good, either may become a force for tremendous evil. It seems to me that because of this they should work together to prevent this evil from happening as it has so often in the past and continues, even now, at present in various places.

2 What is your view on the origin of the universe: both on a scientific level and—if you see the need—on a metaphysical level?

This question assumes a fact that any scientist will immediately argue is unsupported—namely, that there is an origin to the universe. Therefore I must answer the question somewhat differently. My present view of the universe is that it is like an infinite foam of bubbles, each bubble being a universe like or unlike our own, expanding like our own or, possibly, contracting to an infinitesimal point, a singularity. There may be, of course, some universes that are neither expanding nor contracting. I also admit the possibility of alternate or parallel universes in different space-times or dimensions.

There are two possibilities for this theorized infinity of universes or, indeed, even for a singular universe: (a) They (it) always existed and they (it) may or may not always exist in the future but in either case they have (it has) no origin. (b) They were (it was) created at some time in the past and they (it) may or may not continue to exist forever. Possibility (b), of course, assumes a creator. For possibility (a), no creator is necessary even though the concept is difficult.

There is so much in the physical nature of the universe we inhabit, the exact balances of everything needed to support life, the piling of coincidence on coincidence, every one of which is vitally necessary for development of a stable star with a planet that can support life. Even the coincidences that must occur for this planet to have oceans with moderate temperatures over four billion years are highly unlikely. These physical properties of the universe lead me to favor a Designer or Creator, and my answer to Question 3 reinforces this view.

3 What is your view on the origin of life: both on a scientific level and—if
you see the need—on a metaphysical level?

To understand the development of my views, it is necessary to
briefly consider that I have been studying and teaching biochemistry (molecular biology) for over fifty years. The first book I used
to teach biochemistry with was published in 1936, and it contained a significant fraction of what was known in the field at that
time. Today it would take a very large library indeed to contain
all the texts, journals, and other types of publications in these
fields and related areas of biology, chemistry, and medicine. As
time went on, layer upon layer of complication was added that
even today appears not to have leveled off. If one considers even
a single protein, for example, glycogen phosphorylase, this dis-
plays such an immense complexity that it boggles the mind.
Considering the processes of protein synthesis, DNA replication
and repair, and hundreds of equally complicated processes, one
is left with a feeling best described as awe.

I have carefully studied molecular, biological, and chemical
ideas of the origin of life and read all the books and papers I could
find. Never have I found any explanation that was satisfactory to
me. The basic problem is with the original template (be it DNA or
RNA) that would have been necessary to initiate the first living
system that could then undergo biological evolution. Even
reduced to the barest essentials, this template must have been
very complex indeed. For this template and this template alone,
it appears it is reasonable at present to suggest the possibility of
a creator.

I must be honest and point out that even though, as some
have calculated, the odds of such a template forming by chance
are 1 in about 10^{300} or, possibly, a much larger number, it is always
possible to win the lottery the first time you play it. So I agree that
chance is a possibility, but I favor a creator for this template.
There is, as has already been noted, much in the physical nature
of the universe that is so unusual that it supports this view.

4 What is your view on the origin of *Homo sapiens*?

The apparent sequence of evolutionary development of *Homo
sapiens* is slowly being worked out with great difficulty and with
regular changes as new fossils are found. All the published
sequences place *H. sapiens* on a branch that separated some millions of years ago from the chimpanzee and gorilla who are close

relatives. Thus, despite the connotations, unpopular to some persons, I believe we evolved from the primates by a series of fortuitous changes which are unknown. Even when it is theorized what some of these changes are, it is not always easy to explain how they came about. Among these changes could be bipedal locomotion, increased ability to vocalize a variety of sounds, language, use of tools, and agriculture, not necessarily in order of occurrence.

I do not believe that it is inevitable that once evolution began, the certainty was that sentient beings would evolve. The odds for this happening must, again, be very great. Without the extinction that killed most of the reptiles sixty-five million years ago, an intelligent lizard might have developed, or without land surfaces, an intelligent aquatic animal.

I believe further that *H. sapiens* is still evolving rapidly and will continue to do so for the foreseeable future.

5 How should science—and the scientist—approach origin questions, specifically the origin of the universe and the origin of life?

There is only one way to proceed, as we have been doing: to theorize, observe, and experiment. The recent success of the stellar physicists in confirming predictions about supernova mechanics with SN 1987 A is only one example. Building of the supercollider/superaccelerator will hopefully give additional insight on the early history of the universe.

As far as the origin of life is concerned we need more experimentation to help us understand chemical evolution and the first self-replicating organism or cell. Recent experiments on self-splicing, catalytic properties and binding of L-amino acids by RNA have opened a fertile field for this. We also need to have a space probe which can explore the universe and report on life, if any, on other planets. We could almost stretch our technology to do this now, but it should be easily possible in another hundred years. We need to know whether life, either simple or sentient, is widely prevalent, rare, or non-existent throughout our universe. If widely prevalent, this would suggest that a template or templates were widely scattered by an intelligent designer. If we are indeed alone this would suggest the possibility of a lucky chance event, although intelligent design would still be an option.

6 Many prominent scientists—including Darwin, Einstein, and Planck—have considered the concept of God very seriously. What are your thoughts on the concept of God and on the existence of God?

> I have thought intently about the concept of God. For most people, as already noted, it is extremely useful. It is comforting to believe that there is purpose in the universe and in our lives. It would be comforting to know that some deity has an interest in the human race. It would be comforting to know that there is the possibility of life after death. However, there is no hard scientific evidence for a creator. We can suggest, as I have, that the miracle of the initial origin of life and its present complexity is circumstantial evidence. We can suggest, as others have, that the astounding balance of physical properties in our universe that makes life possible is circumstantial evidence, but as a scientist I find these arguments not wholly satisfactory—not much more than a reasoned guess.
>
> I am interested in the nature of Jesus and his counterparts in other religions. Can one really believe the miracles that have been recounted, or does one seek naturalistic explanations? At the least one must attribute extraordinary mental powers to these saviors. And then what really happened in that field in Fatima, Portugal, in 1907 that was witnessed by 70,000 people? Was it just a rare meteorological phenomenon? As science advances the occurrence of miracles that are widely witnessed seems to have decreased. This may be significant.
>
> The existence of a deity may be unknowable unless the deity chooses to let us know unequivocally. Thus far, if a deity exists, this has not happened. Nevertheless, for the circumstantial reasons stated, I believe in a creator. I do not believe that the creator is omnipotent or omniscient, although it may be possible for the deity to be so.

23

Life, Even in Bacteria, Is Too Complex to Have Occurred by Chance

• • • • • • • • • •

Professor Harry Rubin

• Born 23 June 1926

• D.V.M., Cornell University, 1947

• Currently Professor of Molecular Biology and Research Virologist to the Virus Laboratory, University of California, Berkeley; received the Rosenthal Award, AAAS, 1959; the Eli Lilly Award, 1961; the Lasker Award, 1964; the Modern Medicine Distinguished Achievement Award, 1967

• Areas of specialization: cell growth regulation and malignancy

• Works include numerous scientific papers in scientific journals

• Professor Rubin on:

the origin of the universe: "I accept current views".

the origin of life and of *Homo sapiens*: "Life, even in bacteria, is too complex to have occurred by chance".

God: "The Kabbalists put the problem in the form of the 'Ein Sof', or 'no end to the quest' ".

• • • • • • • • • •

1 What do you think should be the relationship between religion and science?

I do not believe there should be any formal relationship between religion and science. I believe, of course, all scientists should be free to practice any religion of their choice.

2 What is your view on the origin of the universe: both on a scientific level and—if you see the need—on a metaphysical level?

I accept current views, although with a certain degree of skepticism. On a different level, I find metaphorical and midrashic interpretations of the Biblical story of creation to have validity in relations among men and between them and God.

3 What is your view on the origin of life: both on a scientific level and—if you see the need—on a metaphysical level?

> Life, even in bacteria, is too complex to have occurred by chance. I believe it was "created" in the sense that Elsasser defines creativity in his recent book, *Reflections on a Theory of Organisms* (Frelighsburg, Canada: Editions Orbis Publishing). This is not a literal interpretation of the Bible story, in other words, it occurred perhaps billions of years ago. Applied here, creation in Elsasser's sense means the appearance of hereditary novelty that is not mechanistically traceable. It accepts evolution but not the Darwinian mechanisms such as natural selection or gradual accumulations of changes in DNA.

4 What is your view on the origin of *Homo sapiens*?

> Same answer as to Question 3.

5 How should science—and the scientist—approach origin questions, specifically the origin of the universe and the origin of life?

> With humility. Biology in particular needs to be open to an entirely new, non-reductionist set of basic abstractions such as Elsasser has formulated.

6 Many prominent scientists—including Darwin, Einstein, and Planck—have considered the concept of God very seriously. What are your thoughts on the concept of God and on the existence of God?

> The conceptualization of God is even further beyond our capacity than visualizing the particle-wave duality of physics. The Kabbalists put the problem in the form of the "Ein Sof", or "no end to the quest".

24

Religion Is a Concern of the Human Spirit

· · · · · · · · · ·
Professor H. G. Schlegel

- Born 24 October 1924
- Ph.D., University of Halle/Saale, 1950
- Chairman of the Institute for Microbiology, University of Göttingen, since 1958
- Areas of specialization and accomplishments: physiology and biochemistry of chemosynthetic and photosynthetic bacteria of soil and water; regulation mechanisms in autotrophic bacteria; works include *Anreicherungskultur und Mutantenausleses*, 1965; *General Microbiology*, 1969; editor of various symposium volumes, handbooks, and of *Archives of Microbiology*

· · · · · · · · · ·

I answer your letter of January 1989, which concerns the oldest and at the same time current questions of thoughtful people. I answer the letter neither as president of the academy of science in Göttingen nor as a microbiologist but as a human being who intends to serve his public.

You are right: In the treatment of fundamental questions, such as the existence of the universe and the creation of life, we are led to religion. The question of the relation between religion and science at the age of twenty, when I returned home from the Second World War, interested me more than it does today. At the time I read the writings of Max Hartmann, Nicolai Hartmann, Max Planck, E. Schrödinger, P. Jordan, Bernhard Bavink, and, of course, the Bible, in order to acquire a personal judgment concerning wrong and right.

Today I see the religious question more from the practical side. Religion is a concern of the human spirit. It strives to accept a mystic world foundation. Without the spirit these problems would not exist. The main question is a different one for me: Is religion an innate necessity of man? Hence the question whether the intellectually disposed, the spiritually minded people of our time, among them the geniuses, should do something, in order to contradict the turning away from religion, the departure from churches toward atheism. According to my view we should do this. Without accepting a new proof of God's existence, I would like to say that I assume that religion belongs to the fundamental necessity of man.

25

In a Scientific Sense, We Know Very Little on the Origin of Life

· · · · · · · · · ·

Professor Robert Shapiro

- Born 28 November 1935

- Ph.D. in chemistry, Harvard University, 1959

- Currently Professor of Chemistry, New York University

- Areas of specialization and accomplishments: chemistry of nucleic acids; chemical mutagenesis and carcinogenesis; heterocyclic compounds; works include (with Gerald Feinberg) *Life Beyond Earth*, 1980; *Origins: A Skeptic's Guide to the Creation of Life on Earth*, 1987; over 80 scientific papers

- Professor Shaprio on:

  the origin of the universe: ". . . I accept the theories of those who work in this area".

  the origin of life: "In a scientific sense, we know very little [on the origin of life]: there has been little opportunity to gather relevant data".

  the origin of *Homo sapiens:* "Humans arose over a number of million years by gradual evolution from a common ancestor we share with other primates".

  God: "I have no reason to exclude the existence of a higher being a priori", but "at present there is no evidence that impresses me".

· · · · · · · · · ·

1 What do you think should be the relationship between religion and science?

I view science and religion as separate disciplines, sharing the attribute that both are conducted by humans. Each seems to serve an important function for its followers.

2 What is your view on the origin of the universe: both on a scientific level and—if you see the need—on a metaphysical level?

I have no formal or personal religion, so I must rely on science for an answer. This is not my area of expertise, so I accept the theories of those who work in this area: the Big Bang, and so forth.

3 What is your view on the origin of life: both on a scientific level and—if you see the need—on a metaphysical level?

This *is* an area with which I am quite familiar. As a scientist, I have no taste for supernatural explanations. In a scientific sense, we know very little: there has been little opportunity to gather relevant data. My best guess is that life arose on earth somewhere between 3.5 and 4 billion years ago. The early steps involved some form of chemical self-organization, driven by available free energy. The specific chemical cyles and energy source involved remain unknown, but it is unlikely that the prominent large molecules of life today, such as nucleic acids, played any role in this process.

4 What is your view on the origin of *Homo sapiens*?

See Question 2. Again, I can only endorse the best ideas of the scientists who work in the area. Humans arose over a number of million years by gradual evolution from a common ancestor we share with other primates. Modern man appeared within the last several hundred thousand years.

5 How should science—and the scientist—approach origin questions, specifically the origin of the universe and the origin of life?

On the origin of the universe: again, it is not my area. Those in the field seem to be making reasonable progress using the available tools (though the supercollider would help).

A certain amount will be learned about the later steps in the origin of life by analysis of the DNA sequence data that is accumulating rapidly. Efforts could also be made in the laboratory to study the early steps of a chemcial evolution process. The so-called "pre-biotic" experiments run thus far have, in my opinion, taught us little. This is discussed at length in my own book, *Origins*.

The most promising source of information on the origin of life is unfortunately the most expensive one: further exploration of our solar sytem and beyond in the hope of observing the process directly.

6 Many prominent scientists—including Darwin, Einstein, and Planck—have considered the concept of God very seriously. What are your thoughts on the concept of God and on the existence of God?

> I have had no personal revelations or private sources of information to convince me of the existence of God. I have no reason to exclude the existence of a higher being *a priori*. If one chose to reveal himself unambiguously and repeatedly to humanity, for example, I could certainly be convinced. At present there is no evidence that impresses me. I hear only a confusion of voices, each claiming to have the truth and denouncing those who make the same claim but arrive at different conclusions.

26

I Do Not See How Science Can Shed Light on the Origins of Design

· · · · · · · · · ·

Professor George D. Snell

- Born 19 December 1903

- Ph.D. in genetics, Harvard University, 1930; Nobel Prize for Physiology/Medicine (shared with Baruj Benacerraf and Jean Dausset), 1980; received the Nobel Prize with Benacerraf and Dausset "for their discoveries concerning genetically determined structures on the cell surface that regulate immunological reactions"

- Currently Senior Staff Scientist at the Jackson Laboratory, Bar Harbor, Maine

- Works include *Cell Surface Antigens,* 1973

- Professor Snell on three areas where science fails to reveal the ultimate truth:

—". . . matter and consciousness appear to me to be distinct though tightly linked entities"; science can tell us "essentially nothing as to the nature of consciousness".

—"While science presumably will continue to provide evidence relevant to the existence or non-existence of design, I do not see how it can shed light on the origins of design, if such there be".

—the matter of first causes: "When I try to explore pathways leading into this area, I find they all come to dead ends".

· · · · · · · · · ·

What do you think should be the relationship between religion and science?

In trying to answer this question, it will be well to start with an examination of the nature of religion and of science.

Religions typically consist of two parts, a set of dogmas or beliefs and a moral code. Dogmas or beliefs are statements of fact or presumed facts. The appropriate question to ask about them is whether they are probable or true. Moral codes, on the other hand, are assemblages of rules, statements of what we "ought" to do. The appropriate question to ask about rules is the consequences of observing or not observing them. Questions concerning their truth are irrelevant.

Science is a refined methodology for generating bodies of facts or knowledge. This makes it directly relevant to the first part of religion, its dogmas or beliefs. It is an appropriate, in fact the only appropriate, tool for testing these, though this does not necessarily mean that it can always provide answers. Science, on the other hand, cannot tell us what we ought to do, but it does have relevance to moral law in that it can provide information as to the consequences of our acts. For example, science can warn us of policies or practices that will harm either the environment or the social fabric. The warnings of science concerning the causes and dangers of acid rain are an example. I am a firm believer in using science and its component, reason, in this capacity, and have discussed this area in *Search for a Rational Ethic*, published by Springer in 1988. I will not elaborate further here on this relation of science to the moral aspects of religion.

Any attempt to provide proof or disproof of religious beliefs would be a lengthy undertaking. I shall not undertake it here. What may be of some relevance is to comment on my personal views on the matter.

Let me say first that because many people, including some good friends of mine, find comfort in the dogmas of religion, I normally keep my disbelief very much to myself. I make an exception in this context, however.

While in general I am a disbeliever, I do find it necessary to note a few qualifications. I shall confine my discussions to these qualifications.

I am, first, a dualist in the sense that matter and consciousness appear to me to be distinct though tightly linked entities. Consciousness we know directly only through our personal, inner sensations, but as I see it, and as I assume other people see it, it is quite unlike the material world. Yet evidence from a variety of sources, including twin studies which show that identical brains produce identical thought and feelings, indicate that consciousness is entirely under control of the brain, and hence of matter. The only possible exceptions are presumed cases of extrasensory perception, whose reality I am not prepared to totally deny, though there are many competent skeptics. While science can study the nature of the material world and is the appropriate tool for the study of extrasensory perception, it can tell us essentially nothing as to the nature of consciousness.

My second qualification concerns design in the universe. I recently have been devoting some time to an examination of our very special world, including those features, both material and astronomical, which make it such a wonderful habitat for life. I seem to see some properties of matter, life, and our world that suggest design. I fully realize that this is a very tricky subject, and I am unwilling to take any firm position, but the fact remains that I am impressed with the apparent evidence that reality is not pure accident. While science presumably will continue to provide evidence relevant to the existence or non-existence of design, I do not see how it can shed light on the origins of design, if such there be.

The third area where science perhaps fails to reveal the ultimate truth is in the matter of first causes. When I try to explore pathways leading into this area, I find that they all come to dead ends. I just can't reach any conclusions.

If and insofar as these are aspects of reality which science cannot probe, this would seem to leave an area of reality or human experience which might be regarded as of religious significance. These areas, as I see them, however, are of a far more specialized nature than the typical dogmas of religion.

These few observations in regard to the inability of science to deal with all areas of human experience and the human environment certainly would leave me a heathen in the view of most religious people, so long as I do not accept their religious beliefs, but they also must brand me as unscientific in the eyes of many scientists, including probably some of my good friends.

27

A Deeper Connectivity than the Mechanical Models of Our Current World View May Comprehend

• • • • • • • • • •

Professor Jeffrey I. Steinfeld

- Born 2 July 1940

- Ph.D. in chemistry, Harvard University, 1965

- Currently Professor of Chemistry, Massachusetts Institute of Technology

- Areas of specialization and accomplishments: molecular spectroscopy; energy transfer in molecular collisions; applications of lasers to chemical kinetics and atomspheric monitoring; works include *Molecules and Radiation*, 1974; edited *Electronic Transition Lasers*, 1976; *Electronic Transition Lasers II*, 1977; *Laser and Coherence Methods in Spectroscopy*, 1977; and *Laser-induced Chemical Processes*, 1981; about 141 scientific papers in scientific journals

• • • • • • • • • •

My own professional pursuits have been in the rather pragmatic and empirical field of physical chemistry. Thus, I cannot pretend expertise in the scientific investigation of the profound philosophical questions reviewed, for example, by V. Weisskopf in his article in *American Scientist* 71, 473 (1983). I have, however, attempted to follow accounts of developments in this field. Perhaps it reflects my own intellectual limitations, but I have become convinced that, at some level, physical reality must be more complex than our conscious minds are able to comprehend. Is it not possible that most (or perhaps all!) of the currently fashionable cosmological constructs, such as "superstrings", "folded dimensions", "wormholes", "pocket universes", and so on, might be just as fanciful as the cosmology of medieval scholasticism, with its angels, crystal spheres, and *primum mobile*?

To say that certain phenomena lie beyond our comprehension might be regarded as a statement of despair from a scientist; on the contrary, I find it to be a source of reassurance, and even inspiration. For in this complexity may lie the potential for a deeper "connectivity" of physical objects, including our organic selves, than the mechanical models of our current world view may comprehend. Since, according to the standard inflationary model, or the somewhat inelegantly named "Big Bang" theory, all of physical reality appears to have developed from a single, uniform, homogeneous entity, could there not be interactions among disparate physical objects which transcend the force laws and the space-time dimensions that our minds seem to be capable of analyzing? This is not in any way meant to suggest that space-time events could be subject to seemingly capricious intervention from "miraculous" forces beyond our comprehension; rather, as the poet tells us, "There are more things in heaven and earth . . . than are dreamt of in your philosophy".

The spiritual impulses that characterize us as human beings may be, in some way, an attempt to recognize these underlying connections, even though our minds are incapable of comprehending them.

28

The Existence of Some Creative Impulse at the Very Beginning

· · · · · · · · · · ·

Professor János Szentágothai

- Born 31 October 1912

- M.D., University of Budapest, 1936

- Currently Professor of Anatomy, Semmelweiss University Medical School, Budapest, and Emeritus President and Research Professor of the Hungarian Academy of Sciences

- Areas of specialization and accomplishments: research on nerve junctions to elucidate circuitry and connectivity in central nervous system; spinal cord pathways and their connections; structure of simplest reflex paths, pathways, and mechanisms of labyrinthine eye-movement reflexes; visual system relay mechanisms; structural basis of nervous inhibition in general; cerebellar pathways and neuronal machinery of cerebellar cortex; anatomical bases of nervous control of endocrine functions; was one of the first to apply successfully the method of experimental degeneration to the tracing of nervous pathways up to their synaptic endings; works include *Die Rolle der Einselnen Labyrinthrezeptorenj Bei der Orientation von Augen und Kopf in Raume*, 1952

- Professor Szentágothai on:

  the origin of the universe: ". . . emergent knowledge about the origin of the universe—whatever its future development—will never be able to disprove the existence of some creative impulse at the very beginning".

  the origin of life: ". . . the first raw materials of life were amino acids synthesized by electric discharges in the primeval atmosphere . . ."

  the origin of *Homo sapiens*: ". . . man arose from some anthropoid ape ancestor . . ."

  God: ". . . I am fully convinced of the existence of God".

· · · · · · · · · · ·

1 What do you think should be the relationship between religion and science?

> There is, in my humble judgement, no direct relation between religion and science. Science cannot even raise, not to speak of answering, questions that form the core of religious beliefs, such as: What is the meaning (or purpose) of the universe, of life in general, and of our existence as conscious beings in particular? It seems silly, therefore, to look for proofs of God's existence or non-existence. This quest would be particularly meaningless for Christianity (and Judaism) in which a God as characterized in the Scriptures would be uninterested in the worship by living beings created "in his own image", if there were obvious direct proofs of his existence. Religion, though, may be very relevant in giving a system of values for man, to be used as guidelines for his behavior under the changing circumstances of history. (This does not mean to deny the fact that comparable value systems can be created and adhered to by radical atheists.)

2 What is your view on the origin of the universe: both on a scientific level and—if you see the need—on a metaphysical level?

> As a humble neurobiologist I am, of course, unable to judge the merits of or recognize the flaws in modern theories of cosmology. However, I do think that we are here on the wrong track and that emergent knowledge about the origin of the universe—whatever its future development—will never be able to disprove the existence of some creative impulse at the very beginning.

3 What is your view on the origin of life: both on a scientific level and—if you see the need—on a metaphysical level?

> I do think that present day theories about the origin of life on this planet are probably correct in assuming that the first raw materials of life were amino acids synthesized by electric discharges (or possible alternative mechanisms of chemical synthesis) in the primeval atmosphere and were gradually enriched in smaller bodies (ponds) of surface water, and their interactions were promoted by adsorption of local aluminium silicate minerals. I do think that Manfred Eigen's hypercycle theory might indicate how the early chemical constituents became self-reproductive. The emergence of the first eukaryotic cell may have taken over a billion years by symbiosis of several prokaryotic organisms (primordial cells, mitochondria, chloroplasts, cilia). I would not give

much chance to hypotheses relying on extraterrestrial sources of life, neither do I think much of so-called "exobiology": the conditions for emergence of life may exist on many other planets of our universe; however, reckoning with the distances and the life spans of human beings I would consider such hypothesizing as "science fiction", if not an outright nuisance.

4 What is your view on the origin of *Homo sapiens*?

There is no question whatever in my mind that man arose from some anthropoid ape ancestor and that many different types of hominids were living mainly in Africa, and probably also in large areas of the Euro-Asian continent. The close genetic and biochemical relationship between chimpanzee and human, and also many hominids in Africa forcefully suggest this continent as the cradle of man. I am not so confident about man's direct descent from *Australopithecus* because such a drastic reduction of the facial skeleton over so short time is difficult to imagine. Discoveries in Europe (Hungary's *Rudepithecus*) and in Asia of different kinds of *Ramapithecus*, might indicate that the spectacular reduction of the facial skeleton may have been programmed into the evolution of the prehominid groups. We are probably still very far from being able to make reasonable assumptions from which point of development prehominids might be considered really human, or even to agree on criteria on the basis of which such a decision could ever be made. There is little likelihood that we shall have a final answer even in the very distant future.

5 How should science—and the scientist—approach origin questions, specifically the origin of the universe and the origin of life?

I am quite satisfied with the development of our understanding as it is, and would like to live for another hundred years only to see how the development of our insight will go on. However, being seventy-seven years of age, I have to be content with what science has achieved until now.

6 Many prominent scientists—including Darwin, Einstein, and Planck—have considered the concept of God very seriously. What are your thoughts on the concept of God and on the existence of God?

As an active (protestant) Christian I am fully convinced of the existence of God, and of everything so succinctly and beautifully phrased in our apostolic credo. However, as a scientist, I am appalled by the emergence and worldwide takeover by irrational religious fundamentalisms. If these attitudes gains territory in our European-American civilization, which has after all created contemporary science and technology, there is indeed much to worry about. Fundamentalism and intolerance are twin brothers and dangerous ones for societies that have the means at their disposal first, to enslave whole populations both economically and spiritually (as amply demonstrated by the Eastern-Communist [Stalinist], experiment or you might say with equal justification the Hitler-Nazi experiment), and second, to wipe out our entire civilization by an atomic holocaust.

It is indeed deplorable that significant segments of Western societies (or people) would rather stick to the anthropomorphistic concept of the Creator as an old, bearded, jealous, and vengeful father—an idea logical to men some five to three thousand years ago—who formed man six thousand-odd years ago from a piece of clay, instead of accepting, with gratitude and admiration for the Creator, our present scientific insights. Or with what do the religious fundamentalists think the Holy Spirit ought to have inspired the writer of Genesis? Perhaps, to anticipate our present-day scientific cosmology and our theories about the evolution of life? I do also believe that the Holy Spirit has inspired the writers of both the Old and of the New Testament, not the words though, but the essence of the message. You may look at the essential part of the message from whatever viewpoint you like: for the modern understanding of man's nature and lost paradise through the development of our brains and its ultimate consequence that we are no more equal to our biological nature and have, hence, to rely on values as guides for our choice between good and evil is indeed the great ultimate truth. The biblical story of man's fall from grace is the most wonderful parable that tells us—or at least tells me—everything that we need to know about our human condition. I could go along for volumes to demonstrate from the Holy Scripture that it is this attitude with grateful acceptance of the Creator's truth and mercy—as so very clearly articulated by the late Donald M. MacKay—that there is simply no other way for educated people at the end of the twentieth century. Alternatively, we could become atheists, but this hypothesis has to be believed, too.

29

Life and Mind in the Universe

• • • • • • • • • •

Professor George Wald

• Born 18 November 1906

• Ph.D. in biology, Columbia University, 1932; Nobel Prize for Physiology/Medicine (shared with Ragnar Granit and H. Keffer Hartline), 1967; received the Nobel Prize with Granit and Hartline "for their discoveries concerning the primary physiological and chemical visual processes in the eye"

• Higgins Professor of Biology at Harvard University until 1980

• • • • • • • • • •

In my life as scientist I have come upon two major problems which, though rooted in science, though they would occur in this form only to a scientist, project beyond science, and are I think ultimately insoluble as science. That is hardly to be wondered at, since one involves consciousness and the other, cosmology.

The consciousness problem was hardly avoidable by one who has spent most of his life studying mechanisms of vision. We have learned a lot, we hope to learn much more; but none of it touches or even points, however tentatively, in the direction of what it means to see. Our observations in human eyes and nervous systems and in those of frogs are basically much alike. I know that I see; but does a frog see? It reacts to light; so do cameras, garage doors, any number of photoelectric devices. But does it see? Is it aware that it is reacting? There is nothing I can do as a scientist to answer that question, no way that I can identify either the presence or absence of consciousness. I believe consciousness to be a permanent condition that involves all sensation and perception. Consciousness seems to me to be wholly impervious to science. It does not lie as an indigestible element within science, but just the opposite: Science is the highly digestible element within consciousness, which includes science as a limited but beautifully definable territory within the much wider reality of whose existence we are conscious.

The second problem involves the special properties of our universe. Life seems increasingly to be part of the order of nature. We have good reason to believe that we find ourselves in a universe permeated with life, in which life arises inevitably, given enough time, wherever the conditions exist that make it possible. Yet were any one of a number of the physical properties of our universe otherwise—some of them basic, others seemingly trivial, almost accidental—that life, which seems now to be so prevalent, would become impossible, here or anywhere. It takes no great imagination to conceive of other possible universes, each stable and workable in itself, yet lifeless. How is it that, with so many other apparent options, we are in a universe that possesses just that peculiar nexus of properties that breeds life? It has occurred to me lately—I must confess with some shock at first to my scientific sensibilities—that both questions might be brought into some degree of congruence. This is with the assumption that mind, rather than emerging as a late outgrowth in the evolution of life, has existed always as the matrix, the source and condition of physical reality—that the stuff of which physical reality is composed is mind-stuff. It is mind that has composed a physical universe that breeds life, and so eventually evolves creatures that know and create: science-, art-, and technology-making animals. In them the universe begins to know itself. Also, such creatures develop societies and cultures—institutions that present all the essential conditions for evolution by natural selection (variation, inheritance [mainly Larmarckian], competition for survival)—so introducing an evolution of consciousness parallel with, though independent of, anatomical and physiological evolution.

[The preceding is the abstract of Professor Wald's article, "Life and Mind in the Universe", *International Journal of Quantum Chemistry: Quantum Biology Symposium* 11 (1984): 1–15. Copyright © 1984 John Wiley and Sons, Inc.]

30

I Don't See How We Can Gather Empirical Evidence about How the Natural Order Itself Came into Being

· · · · · · · · · ·

Professor Ward B. Watt

- Born 21 October 1940

- Ph.D. in biology, Yale University, 1967

- Currently Professor of Biology, Stanford University

- Areas of specialization: study of adaptive mechanisms and microevolutionary processes in natural insect populations from perspectives of biochemistry, physiology, genetics, and ecology

- Professor Watt on:

  the origin of the universe: "At this time, I don't see how we, as creatures within the natural order, can gather empirical evidence about how the natural order itself came into being".

  the origin of life: ". . . it now seems apparent that the origins of life from prebiotic conditions are understood in principle."

  the origin of *Homo sapiens*: ". . . conscious, reflective intelligence is probably an emergent property of complex living systems anywhere in the universe that these occur".

  God: "I do find the intuitive concept of a 'higher power' who might stand behind, pervade, the whole natural order in its lawful operation, far more impressive than the 'God as obsessive tinkerer' image which fundamentalists of several persuasions purvey".

· · · · · · · · · ·

1 What do you think should be the relationship between religion and science?

"Science" is the human process of seeking to understand how the natural order works; as Jacob Bronowski once remarked, it is simply the Roman word for "knowledge". It proceeds by the test of evidence and the rule of reason, asking questions with imagination and answering them with whatever pertinent evidence can be found. Science explicitly avoids reliance upon "established" human authority or recourse to the "extraordinary agents" which Charles Lyell first proscribed from geology. Science can answer no question in the absence of evidence; and scientists also remember that absence of evidence is not equivalent to evidence of absence.

Religion is, to me, a personal and intuitive system of belief about the ultimate meaning of existence as we know it. The basically deist, yet anti-dogmatic, writings of Thomas Jefferson and of Henry Thoreau on the subject resonate with my own feelings quite closely. It seems such intuition must be constrained or informed by our empirical experience. For example, if there is indeed a Supreme Being, he must either be unmerciful or else self-constrained from intervening in the affairs of the empirical world on a day-to-day basis. No other alternatives are compatible with the routine human experience of unmerited personal disasters, or atrocities such as the Hitlerian holocaust. (Non-intervention seems to me to be pre-condition for the genuine free will of beings within the empirical world, but that's a separate discussion.)

I believe that issues should be resolved in the everyday world by what amounts to the approach of science (the test of evidence) as informed by human values. That doesn't imply that all human values can be derived from science, though Bronowski made a pretty good start at showing that at least some of them *can* be thus derived. But I have no problem with postulating the value of, for example, humane kindness as a primary "given", or primitive concept, on which to base judgments of evidence as to right and wrong. Any system of propositions must begin with the active choice of primitive concepts.

Organized-group religion, though I personally find no need of it, seems to help and comfort many people in a world which they often find threatening and mysterious. It seems to me that it becomes dangerous to human welfare when, and only when, it adopts a claim to exclusive possession of the truth and begins to justify vicious actions toward others, or other disfunctional or evil behavior within society, on the grounds of such claims to exclusive truth.

Organized-group religions are often perverted by power-hungry persons via the assertion of dogmas which allow them to dominate the minds and actions of their followers. This clearly explains the reaction against *any* form of religion often seen in persons most devoted to human dignity and independence.

2 What is your view on the origin of the universe: both on a scientific level and—if you see the need—on a metaphysical level?

> Here is a major question among those which we may be able to formulate but not answer. At this time, I don't see how we, as creatures within the natural order, can gather empirical evidence about how the natural order itself came into being. Some subsidiary aspects of this question are: Why is an electron an electron? Why does the Pauli exclusion principle operate as it does? Why do the chemical elements C, O, N, P, Fe, and so on display their remarkable "suitedness" for the chemical support of living processes, and indeed for the emergence of those processes from simpler beginnings?
>
> These are not meaningless questions in any sensible human terms, though some claim that they are. (Such persons seem to me simply to be narrowing their view of meaning to avoid apparent embarrassment at our ignorance.) Yet I don't see how we can get answers to these questions. On the other hand, the argument from failure of imagination ("I just can't think how that could work!") has been discredited so very many times that the possibility of answering these questions empirically in due course *must* still be recognized.

3 What is your view on the origin of life: both on a scientific level and—if you see the need—on a metaphysical level?

> This question falls within the proper scope of evolutionary biology. Empirical progress on this issue began with the first simulation experiments of Stanley Miller, Philip Abelson, and others in the 1950s, showing that lightning under a variety of early-earth-atmosphere scenarios can form biological monomers in good yield. There has been immense progress since on prebiotic synthesis of a diversity of biological monomers under a variety of conditions, as well as demonstration that such materials form spontaneously in meteoritic materials, comets, and so forth and are ubiquitously distributed through all parts of the universe which we can observe, even by radio astronomy! An encouraging beginning has been made at understanding possible mechanisms of prebiotic polymerization and even on the spontaneous emergence of proto-cellular boundaries between "self" and "non-self" in aqueous solutions of amphiphilic molecules. In short, while there is a great deal of utterly fascinating work yet to do—hard problems yet to solve and much more

evidence to gather—it now seems apparent that the origins of life from pre-biotic conditions are understood in principle. If this view continues to hold up, then we have the remarkable result that life is an intrinsic property of the most fundamental aspects of the physical and chemical order and is certainly widespread throughout the universe wherever planets, enriched for the carbon-through-iron portion of the periodic table, exist at the proper distances from R-, G-, or K-series stars.

4 What is your view on the origin of *Homo sapiens?*

This question also falls within the proper scope of evolutionary biology. Forty years ago the fossil record of our own group of creatures, the hominids and pongids, was very nearly absent except for its most recent representatives. The gap of time over which relevant evidence was absent stretched from *Aegyptopithecus* to *Homo erectus*—say twelve million years or more. Imaginative and energetic effort by palaeontologists has narrowed the largest remaining gap to three to five million years at most and has shown us a great deal about how our physical *and* mental nature has arisen. This evidence, combined with new understanding of our nearest relatives, the great apes, emphasizes that conscious, reflective intelligence is probably an emergent property of complex living systems anywhere in the universe that these occur. It also emphasizes that the mind-body problem is likely also to be an emergent problem and one that at least includes our near relatives, the great apes, as well as ourselves.

5 How should science—and the scientist—approach origin questions, specifically the origin of the universe and the origin of life?

I find the question itself peculiar. Why should the inquiring mind approach any particular class of questions in a way different from the usual? Science should apply a "business as usual" approach to "origin issues": ask imaginative questions and try to gather evidence to test them! It does seem to me that we can ask meaningful questions which we may find difficult, or perhaps ultimately impossible (but we should be slow to conclude that!), to answer.

6 Many prominent scientists—including Darwin, Einstein, and Planck—have considered the concept of God very seriously. What are your thoughts on the concept of God and on the existence of God?

I've remarked already that the personally merciful, interventionist God-concept of some fundamentalist religions (let alone the animist/polytheist pantheons of other cultures) seems to me to be ruled out by ordinary human experience. It seems to me that most working scientists have a profound sense of awe at the overarching majesty of the universe, as we are coming to know bits and pieces of it better and better. Some, perhaps in reaction to the manipulativeness of "professional religionists" as I've suggested, recoil from any expression of this awe in terms beyond what we can observe directly in the empirical world. I, personally, do not. I do find the intuitive concept of a "higher power" who might stand behind, or pervade, the whole natural order in its lawful operation, far more impressive than the "God as obsessive tinkerer" image which fundamentalists of several persuasions purvey.

In accord with the idea of limitations on our power to look outside the natural order by empirical means, I would be very surprised if I, or anyone else, could be more clear about the matter than this.

PART THREE

· · · · · · · · · ·

The Existence of God and
the Origin of the Universe:
A Debate between an Atheist
and a Theist

The Existence of the Universe as a Pointer to the Existence of God

Professor H. D. Lewis

Former Head of the Department of the History and Philosophy of Religion, University of London. Chairman of the Royal Institute of Philosophy and first President of the International Society for Metaphysics. Former president of the Mind Association and of the Aristotelian Society. Delivered the Gifford Lectures in 1966–67 and 1967–68. Works include *Philosophy of Religion, The Elusive Mind,* and *The Self and Immortality.*

Let me begin with something that may appear elementary, namely that we and other people and things exist now. I do not see how anyone can doubt that. There will be some widely differing interpretations of what this involves. For some, to say that I exist is to refer to a subject, abiding for some while at least, who is distinct from all his experiences—at the center of them as we say—though presumably dependent on having experiences of some kind. I am, on this view, more than all I am aware of or think, all that goes on in my mental life, the me or self who has this life, the subject in all experience. I am myself inclined to this view, and it is a view that has a distinguished ancestry. Others will say that a person is no more than the patterns of his experiences, that there is nothing strictly beyond the passing states. Others will insist on at least some dependence on the body, and some reject the finality of the distinction between one's mind and one's body; yet others regard the person as just the brain and the body, identifying us with our bodies, often of late in an outright physicalist sense, though those who are bold enough to take that line, without some stronger concession to our consciousness, appear to be a diminishing number at present.

There are these views and variations upon them. But I am not intending to discuss any of the possibilities just noted. Without going into these, and the like, issues, we can quite properly, I think, affirm that in some sense we and other people exist—along with the world around us. The latter admits also of varied interpretations, ranging from a Berkeleyan view that all we perceive in the world around us exists in being so perceived, there being such vast varieties in all our perceptions, depending on the light, distance, and our own organs of perception, nervous systems, and brains, and so on. Others favor one of the many forms of more realist views, including the view that the real world is itself other than what we directly perceive. I do not discuss these, for whatever of such theories we favor, we shall all, I submit, agree that persons or people and other creatures exist, and that they do this in a world that abides in a way familiar to all. We can agree on that, I hope, and let other more difficult and basic philosophical questions wait their turn for the time being. The news in the papers would have much the same significance for Berkeley or Bradley or G. E. Moore.

This is not to trivialize the profound philosophical issues on which such thinkers differ. They are of great importance, in themselves and in further implications. It is not easy for any thinking person to avoid them. But I want to set them aside for my purpose and concentrate, as a starting point of what I wish to go on to, on the fact which we can all admit, that we ourselves and others and the world around us all exist.

We have learned a great deal which most of us accept on the saying of others, such as expert scientists or historians, and as much of what they say as we can understand, about the way all this has come about. We know some things at least about the way this planet was formed, how it solidified and acquired a firm crust over an amazingly long period, very much longer indeed than used to be thought (and still is by those who do not look beyond a literalist understanding of the Bible account of Creation). We know much about the period and conditions in which there came to be some animate existence and the very long period of change and development by which there came to be formed the creaturely life with which we are familiar, the life of animals and eventually of beings like ourselves from some of which we have descended. Even our own development as human beings and the changes in our fortunes which have brought the world to its present state, the periods we think of as genuine history and not "prehistory", seem remarkably long to us, though they are of very slight duration indeed by comparison with the vast ages since the earth on which we live began to be.

This itself is slight in relation to eons of time during which other bodies and systems in the vast physical universe were formed, some of them inevitably far from our own, however we reduce the strangeness of it by counting in light years. The layman understands little directly of these affirmations. He must learn what he can from the evidence of others and digest it as best he may. But no intelligent person today who is familiar even with the little he can absorb from science can fail to marvel and perhaps be overwhelmed at the thought of the amazing periods of time and the varieties and complexities in the course of the vast physical universe as we come to know of it today. This is now a familiar though abidingly amazing story, of which the layman can absorb little beyond the general lineaments of it. Disagreements there may be about points of detail, but none of us will presume to dispute the general story. The universe has been around for a very, very long time. Our period as human beings, though stunningly vast in itself, is infinitesimal by comparison. And no one seriously doubts that. We have learned to take it today, however little our minds can properly comprehend it.

Vast indeed, and beyond proper comprehension. But did it have a beginning? Is there an author and sustainer? Can we meaningfully talk that way any more? This is what I really want to ask, and I want to stress at the start that this is not a question for the scientist as such, but for all. For whatever the scientist tells us, about Big Bangs or whatever seems appropriate, the question I have asked remains independently of any peculiar feature of the story. It is also of first importance.

To those who repudiate the notion that there has to be some original author or sustainer, we may say that they seem to be left with one of two alternatives. One is to say that there must be some sort of beginning but that there is nothing whatsoever to account for it. Some amazingly long time ago, things came to be and shaped themselves into the universe we know, including ourselves. This just happened, out of nothing, a total void. First there was nothing, not even presumably time, and then there began a world or worlds, whatever they were like, some sort of reality at least and processes about which we now have learned a great deal. I suggest that this is a notion we just cannot accept, not because of any religious background or upbringing, but as ordinary sensible beings. *Ex nihilo nihil fit* was said long ago, and seems as unavoidable for us as for those who pondered these things earlier.

There is no precise reason we can give why things should not have started up out of nothing and followed the course they have. But it is not just a hunch that we have that this is impossible. We have to ask, among other things, whether there was time before or not (and it is hard to say there was not at least time itself), why it started just when it did, at a date calculable in principle from now or, alternatively, how we reached now. A sheer fluke (we can hardly say "miracle" in this context), a completely random beginning taking all the same a remarkable subsequent course? From some point just nothing, and then out of the blue, we might put it, genuine realities. It is not just disconcerting for intellectuals to admit themselves quite baffled. It is not just that it is upsetting to have to admit a radical totally chaotic feature of a universe whose course we can otherwise increasingly understand. I can only put it to you that, if you reflect, you just cannot, independently of any further implications, accept the idea of a totally random springing into being out of nothing. Can you avoid asking why the sort of world we find *is*, independently even of its being wonderful in so many ways, if there was simply a random start with no sort of before at all? Why this sort of world, and why did it start when it did? It could be anything from a random start and as readily vanish.

I have said "independently of its being wonderful in so many ways", and I have no concern to withdraw that, but I think it is fair to supplement what I have said with consideration of what the universe, as we ourselves find it, in fact involves. It is a universe in which some amazing intricacies can be understood, so that the scientist can probe into astonishing distances, in time and space, and tell us what to expect and find. Where it is puzzling there are solutions, down to the most elusive procedures in our own bodies. If one falls off a branch one falls to the ground, but far enough into space we are weightless and do not fall down at all. This might have seemed a miracle at one time, or just not believable—no one would venture out of a plane if it got that far. But this is commonplace now, and every schoolboy knows about gravity and being weightless. Not only is all this remarkable in itself, but it is astounding what advances we have made in our understanding of the world around us.

We have become highly intelligent creatures; we have varying degrees of artistic sensitivity, in music, painting, literature, and more; we play, which is a remarkable phenomenon in itself, and some reach high degrees of athletic skills; we have endearments and affections, some deep and sensitive, and high regard for the achievements of others; we have remarkable standards about the way we set things out on paper and in print (some of the most skeptical amongst us, the late Professors Gilbert Ryle and A. J. Ayer, being as exacting in this as any. Can you think of Ryle, as editor of *Mind*, accepting something in shoddy English style or being careless about the way he put things himself?). We have standards of behavior and of obligations to one another; we resent meanness and affectations and have learned to have a profound admiration of high moral attainment or selfless sacrifice, irrespective largely of ethical allegiances; we have a rich social existence and ways of sustaining it, notwithstanding appalling and extensive lapses which leave the religious at least with an exceptionally toughened problem of evil. But I shall not go deeper into these matters at present. Let it suffice to note amazing achievement, which I do not think can be doubted whatever the lapses and failures. For all the suffering, we are a remarkable race; and are we to suppose that after an astoundingly long and complicated process of development, we and other creatures too (in this planet and, so far as we know, nowhere else) have reached the stage we are in and have been for what is for us a long time, out of a beginning, however remote, in nothing? That is just too odd to be credible.

But it may be said, and this brings us to the second alternative, there is no need to bother your head about these troublesome questions of origins or beginnings—for the simple reason that there was no such origin of the world, or the universe, as a whole. There was always something, events of some kind, static or changing and conforming with the same conditions granted changes of conditions or circumstances. However far back you go, there was something, and beyond that and beyond that again *ad infinitum*, without any first point.

This sounds very plausible when we consider the difficulties of fixing, even as a matter of principle without further explication, how there could be such a first point without anything at all to initiate it. But we have difficulties at once over the word 'always'. What is this word doing here? The normal provenance of it is some specific reality and specific conditions, though they may be very extensive: "there is always snow on the top of Everest", "it is always very cold within the Arctic Circle", "it is always warm near the Equator", "the grass always grows in Spring in these climates", "the leaves always fall off the trees in the Autumn", "Williams is always late for school". We can say such things, but "always" remains a limited term. There was a time when there was no snow on Everest, not even an Everest. When we go far enough back the entire world is burning, as most of it is burning now well below the crust. There can be no absolute "always". That is just because we do not know what it means. You can talk of an immensely long way back in time, but *always*— what does that mean? Can we envisage it even in the most general sense possible? And again we have to ask: How do we reach "now"? We seem to have been much led astray by the normal use of "always". It is hard to give it any meaning when extended beyond this.

What, then, do we say? Many philosophers of today (most, I suspect) will have little interest in the question, or they will boldly tell us that we have strayed well beyond the limits of meaningful discourse. Some will dismiss the problem as a pseudo one like many other questions, it is alleged, with which thinkers have bemused and muddled themselves in the past; but if not pseudo, at least pointless. For no intelligible answer is possible, and the best thing to do about our sort of conundrum is to forget about it and get on with manageable questions where there may be some result and profit.

But the question will not go away, however bewildering it proves. The world is real, and so is time, whatever theories we form about them. We would all agree that some things happened long ago, and from that we proceed to yet earlier times, and beyond those. But if we cannot suppose that this goes *ad infinitum*, and if it seems preposterous to assume that the world sprang out of nothing, we do have a genuine problem for philosophers and others to ponder.

One thing is clear. We cannot suppose, as the ground or author or sustainer of all there is, something which remains, in the last account, in essentials of the same sort as the limited, finite things it is to explain. That would still keep us within the same area of discourse. We might invent and postulate something very strange and unusual, or some mind vastly greater than our own, but the question would still remain, "how did that come to be?". We have only extended, if even very vastly, the scope or range of the sort of explanations we give now of things. We have found that the world is vaster and stranger than we thought. But it is still the world (or the universe) as we find it. The ultimate questions stay. It is like the familiar child's question when told that God made the world—"who made God?".

It certainly looks as though we must allow our thoughts to stretch altogether beyond the finite sphere. But how can we do that, since anything we can understand or describe has to be in terms of some limited finite existence, one thing related to others, however odd or unusual? Beyond this there just does not seem to be anything we can say. We reach our limit with the type of explanation we offer of things in the world as they affect one another. Any description is meaningful in these terms, or none. If we try to pass beyond this we can say nothing. What sort of explanation of anything can we offer that is not in finite terms?

This last point seems to me very sound, and we have to be careful, in religious thought and elsewhere, not to forget that. But it does not leave our concern, and our response to it, vacuous. The problem is genuine enough and unavoidable. What we have to say, I submit (and to understand this is not trivial but of immense importance), is that the entire finite world, as we understand it, is rooted in some reality altogether beyond finite existence, an ultimate, and for us irreducible, mystery. But, if we reach our limit with this mystery, if our minds cannot penetrate it in any fashion, how can it be of any significance to us? We might just as well forget it.

The last line is the one many take, and this is understandable; there seem to be more profitable things to do. But this is altogether mistaken. The recognition that all we are, and the world around us, is dependent on some reality altogether "beyond" and not limited or broken up as finite things are, some one whole "transcendent" source of being, is a marvel in itself and affects all our hopes and aspirations. It gives us a conspectus that is not limited or finite. Of itself it does this, though there is more to be said.

Let me put this more simply, if I can, very cautiously. To realize that the world as we find it, our normal milieu, is itself dependent on, or points to, a reality that is essentially complete and enduring, as finite things are not, this makes its own impact in all other awareness, bringing its own new dimension to all. The ultimate is not just subsisting in its own existence beyond, it is all-encompassing, the supreme source and maintenance of all, it lends the wonder of its own existence and wholeness to all. Transcendence, like a rare shining light, is the most notable we can ever know.

This is so all the more because there is an excellent case for saying that the unique completion of the ultimate carries with it essential goodness also. It would not have its proper completeness without that, little though we may know how it operates, as has been impressively shown by Professor Sontag in his *Divine Perfection*. I have not here the time or space to go into this further claim. But the ground of all cannot be less than all or inferior in any regard. This is how we answer those who say there could be many gods, as there well might be if thought of in terms of an ordinary first cause. But we cannot think of a ground of all with limitations. It could not have its unique finality without the appropriate perfection. It does not just happen to be holy, it is essentially holiness.

I stand by this, but I shall have to amplify it elsewhere. There are two other matters to be noted, and just that for the present. In stressing the dependence of all things on their ultimate source, and some immediacy which this involves, we must be very careful to avoid the pitfall, as it seems to me, of treating finite existence as itself some mode or element of the transcendent itself. Various forms of monism do just this. The pattern for most of these in the West was set by Parmenides, although there are milder forms of monism like post-Hegelian idealism in the last century seeking to do more justice to finite existence. In the East we have a more persistent determination to take up the finite altogether into the ultimate, the Supreme Being or Supreme Self, although there are found in Eastern cultures at various times profound and perceptive modes of pluralism. None of these forms of monism seem to me sound. They fail to do justice to the obvious genuineness of limited finite beings and especially the distinctness of persons which is the key to most that we find of worth, pre-eminently moral worth, as I have stressed elsewhere. Each is distinct, uniquely aware of himself, but no one can live wholly for himself or within himself. We need the other as other, both man and God. In multiplicity is our way of life, and our health is in apprehending this. There are genuine and far-reaching claims upon us, and affection is of little worth without the other in its mysterious otherness.

But the ultimate is in one respect radically different. We know this as we appreciate that it is the transcendent ultimate. It cannot be diversified within itself. We cannot properly apprehend this, least of all in its operation on the world it sustains. But we cannot think of transcendent being except as essentially whole. This is the marvel of created, dependent, existence, that it has its limited and essential diversity expressly required by its dependence on an equally essential wholeness. To appreciate this, as in their un-philosophical way the Hebrews did, is the beginning of wisdom in religion and all related matters. Creation is not an illusion but an abidingly perplexing and bewildering reality which, in all its incomprehensibility, we cannot escape.

My final point must be very brief. It concerns the disclosures or revelations which religious people also claim. Religion begins with the transcendent whose life and operation in itself we cannot begin to penetrate. It is, as I have insisted, an all-encompassing but total mystery. But that in no way precludes disclosures within the sort of experience we have. Different versions of this appear at different times and places. It is much affected by what, as finite beings, we already believe, and needs much correction in consequence, as the Bible frequently stresses. The test of revelation, and its imperfection and falsity at times, has been discussed by me elsewhere (in my *Our Experience of God*). At this point the religions of the world differ deeply. We must not blur that or pay the wrong price for amity and esteem. The key here, it seems to me, is in our awareness of one another which is, to my mind, essentially oblique. We do not know others, as we know ourselves directly in having the experiences we do have. We may be very close but we do not have the experiences of even those we love and know best as they are for them. We go by their words and behavior. This, as I have also stressed elsewhere and tried to meet its difficulties, is not to my mind as great as might seem or in any way distressing—in fact the reverse. I have also maintained that our awareness of the transcendent, in all its mysterious elusiveness, extends itself into the total situation in which it happens and thus underlines certain insights which, although valid in their secular soundness, relate especially to our ground and source. In this there is also a patterning and course not unlike our awareness of one another. But this is not the occasion to repeat or amplify what I have maintained elsewhere. I mention it lest anyone suppose that what I have said on the essentially mysterious source of the world exhausts the subject and must be, because of that mystery, our last word. There is indeed much besides at the very heart of religion. There is a "walking with God" and God's giving of himself in one unique person. But these are not matters to be set forth properly now. An allusion must suffice. What I have done has been to set before you the initial claim, as it seems to me (logically at least) of all religion and, above all, to challenge our secular friends to say whether they do not also feel drawn to what I have said earlier and in the body of this paper about finite origins and the like. I am convinced that these are genuine perplexities to which philosophers ought to attend.

Why the Existence of God Is Not Required to Explain the Existence of the Universe

· · · · · · · · · ·

Professor Antony Flew

Former Professor of Philosophy at the University of Reading. Member of the Council of the Royal Institute of Philosophy and the Council of the Freedom Association. One of the best-known expositors of atheism in the English-speaking world, he co-edited the celebrated work, *New Essays in Philosophical Theology* in 1955. Other works include *God and Philosophy*, 1966; *The Presumption of Atheism*, 1976; and *The Logic of Mortality* (forthcoming). Delivered the Gifford Lectures in 1986.

· · · · · · · · · ·

Notwithstanding that the prime duty of the respondent in any academic exchange must be to disagree, nevertheless I begin here by insisting that I am in full accord with Professor Lewis about the proper launching pad for such discussion. Indeed you might say that my agreement is not so much full as actually overflowing! For I would argue, and indeed often have argued, that anyone who can be truly said to understand the skeptical position reached by Descartes in the first two paragraphs of his *Discourse on the Method* must also be in a position to know that any such extreme or—as Hume would have had it—Pyrrhonian skepticism is quite unjustifiable.

In particular I want to go further than Lewis in the extent and confidence of my initial convictions about the sort of creatures which we are. He says: "For some, to say that I exist is to refer to a subject, abiding for some while at least, who is distinct from all his experiences—at the center of them as we say—though presumably dependent on having experiences of some kind. I am, in this view, more than all I am aware of or think, . . . the me or self who has this life, the subject in all experience". Certainly it is preposterous to contend that "a person is no more than the patterns of his experiences, . . . there is nothing . . . beyond the passing states". For the subject of any person's conscious states necessarily is a creature of all too solid flesh and blood; a specimen of the animal species of which all of us are members. Nor does it make any sense to speak of such experiences save as the conscious states of the creature whose states they are. Furthermore—Berkeley to note—if people were not, as we are, material things of a very special sort, then we should each of us be at a loss for believing himself or herself (or itself?) not to be solipsistically alone in the universe.

Disagreement begins when Lewis challenges "those who repudiate the notion that there has to be some original author or sustainer. . . ." We should, I suggest, here follow St. Thomas in distinguishing very sharply between questions about an "original author" and questions about a "sustainer". Thomas, it will perhaps be remembered, insisted in his *de aeternitate mundi contra mumurantes* that, whereas it was possible to prove philosophically that the universe must at all times be totally dependent upon a Sustainer Cause, it was not possible to provide any parallel proof that it had a beginning and was in that beginning created by an uncaused First Cause. Theists would, surely, be well advised to follow Thomas in concentrating upon the former of these two cases.

Certainly, if it is allowed that the universe as we know it did have an explosive beginning—as seems to be the consensus in contemporary cosmology—then the suggestion that it was started off by a quasi- or supra-personal Being, hypothesized to fulfill that function, does have appeal. For, as Thomas Sowell has remarked, "the simplest and most psychologically satisfying explanation of any observed phenomenon is that it happened that way because someone wanted it to happen that way". But that satisfaction should diminish once we recall the extent to which the progress of our individual and collective understandings has consisted in appreciating the extremely narrow scope available for such appealing explanations. At least within this universe only the tiniest proportion of all happenings in fact are the intended results of intended actions and hence truly explicable as such. Nor is it economical to hypothesize an Unstarted Starter in order to explain an apparently unstarted start.

Certainly, if theists can excogitate good reasons for believing that the universe does have a Sustaining Cause, then opponents will find it difficult to deny that, if it did in fact have a beginning, then that beginning must also have been the work of that Cause. It is, however, worth saying, what seems very rarely to be said, that on these assumptions we should surely expect our Infinite Being to be continuously creating an infinity of non-communicating universes; rather than to have created—as a single-shot, one-off operation—only the universe we know.

I doubt, however, that Lewis will be greatly worried by anything which I have said so far. For he insists, and rightly, that "we cannot suppose, as the ground or author or sustainer of all there is, something which remains, in the last account, in essentials of the same sort as the limited, finite things it is to explain. That would still keep us within the same area of discourse". Hence it would very properly provoke "the familiar child's question when told that God made the world—'who made God?'"

But now the trouble, as Lewis also appears ready to recognize, is that this specification must make any discourse about the transcendent reality, to which our universe supposedly somehow points, doubtfully intelligible. To satisfy the proposed job description, the reality has got to be transcendent. Yet, to the extent that we humans are to be able to understand any talk about what is thus hypothesized, it simply cannot be as transcendent as all that.

Even if this difficulty—which is part of what led Hume to conclude that anything behind or before the universe we know must be outwith the range of human understanding—could eventually be overcome, I should still need much more to shift me from what he would have called my "Stratonician atheism". Every series of explanations must necessarily end with some elementary fact or facts, taken to be ultimate and brute. And I do not see how anything within our universe can be either known or reasonably conjectured to be pointing to some transcendent reality behind, above, or beyond. So why not take the existence of that universe and its most fundamental features as the explanatory ultimates?

Response to Flew

· · · · · · · · · ·

Professor H. D. Lewis

I am deeply grateful to Professor Flew for taking trouble to set out quite explicitly and clearly the point at which he is unable to follow my own view. Of course I agree that we have to reckon with the universe being sustained as well as started and had, I fear, assumed that readers would take what I say in that sense. Perhaps I should have been more forthcoming here. But Flew would agree that this issue does not make a radical difference and that I would not be worried by what he says in such respects.

I am pleased also that Flew thinks I go to work in the right way or from "the proper launching pad". The real crux comes when I maintain that the ultimate explanation of things cannot be found within the finite existence with which we are normally familiar, but requires us to postulate some transcendent reality. But the transcendent, by its very nature as we postulate it, can convey nothing to us, it is "doubtfully intelligible" since everything we can understand must be in terms of finite existences and "ends with some elementary fact or facts, taken to be ultimate and brute".

I go along with much of this and have myself complained of those who ease things for themselves by bringing the transcendent too conveniently within the world of finite realities. We have no notion what it is like to be transcendent or how it operates, and much in holy scriptures has stressed the same. But in spite of this I do not agree that it is pointless. It brings a wholly new significance or dimension to all our thought and experience. The task of holy scriptures is largely to bring this out, but I have maintained at some length elsewhere that the impenetrability of the transcendent in itself to us does not preclude our apprehending much in mediated ways about the attitudes and work of the transcendent in relation to us, all the more wonderfully since it has the alleged beyondness in itself. It is God that we distinctly know in these ways.

There is a peculiarly helpful analogy to which we may turn at this point. It is between our knowledge of God and our knowledge of one another. We know much about other persons, and there are some whom we know very closely. But we do not know any other person in the way we know ourselves. We know our own experience in having it, and we know ourselves as the subjects having those experiences. At one point Flew comes very close to admitting all this. It is "preposterous" to contend "that a person is no more than the patterns of his experiences". There is a "subject of any person's conscious states". I have contended much to the same effect, but alas, Professor Flew holds that this obvious subject "necessarily is a creature of all too solid flesh and blood".

I have opposed the last view so extensively already (in the series based on my Gifford Lectures especially), that it would be idle to repeat here all I have said at length elsewhere. I do not see how the subject of consciousness or experience can be other than a severely mental existent, dependent though it may be in some respects at present on material or bodily conditions. The nature of mental awareness or thought plays its part as well, quite obviously it seems to me. But we do not know other subjects (or their experiences) as we know ourselves. Each one is unique as subject in his own case. We have no similar knowledge of other persons, as I have much stressed; we ascribe to them feelings, hopes, disappointments, perceptions, and so on, such as we have ourselves, but we have no notion, even in the closest intimacies, what it is to be the other person to whom all this happens or which he "lives through" (this phrase, in this use of it, we owe to Samuel Alexander; he also spoke in the same way of "enjoying" our mental states) as the unique individual he is.

It is wonderful how all this happens, but since it clearly does, through observation of the behavior, words, and so forth, of other persons and constitutes much of the richness of personal relations, it is less strange that the like may happen, in very different but no less acceptable ways in our relations, with God. Overconfidence may sometimes lead us to think that we know God better than we do or in further ways than we do. But we have, in all humility, to accept the strange situation which brings so much wealth and excitement to all aspects of existence, and we must work hard at the evidence for it, in meditation and living close to established sources. It is mediated and never as easy as we may suppose, but it is also most remarkable.

Response to Lewis

· · · · · · · · · ·

Professor Antony Flew

Notoriously, confession is good for the soul. I will therefore begin by confessing that the Stratonician atheist has to be embarrassed by the contemporary cosmological consensus. For it seems that the cosmologists are providing a scientific proof of what St. Thomas contended could not be proved philosophically; namely, that the universe had a beginning. So long as the universe can be comfortably thought of as being not only without end but also without beginning, it remains easy to urge that its brute existence, and whatever are found to be its most fundamental features, should be accepted as the explanatory ultimates.

Although I believe that it remains still correct, it certainly is neither easy nor comfortable to maintain this position in the face of the Big Bang story. For, apparently, our cognitive predicament is as Lewis describes it in his original paper: "Some amazingly long time ago, things came to be and shaped themselves into the universe we know, including ourselves. This just happened, out of nothing, a total void. First there was nothing, not even presumably time and then there began a world or worlds . . . and processes about which we now have learned a great deal".

This, Lewis continues, "is a notion we just cannot accept. . . . *Ex nihilo nihil fit* was said long ago, and seems as unavoidable for us as for those who pondered these things earlier". But what Hume called "that impious maxim of ancient philosophy" is no logically necessary truth. Our warrant for accepting it either as a methodological maxim or as an a posteriori truism is, and can only be, our experience of and within the universe. It is, surely, the weight of all this experience which makes even those of us who see no need to postulate a sustaining cause for the universe so uneasy about saying that its initial Big Bang had either an unknown cause or no cause at all.

Yet, if and so long as that is the furthest back into the past that scientific investigation is able to take us, is there anything more or other than this which we are rationally entitled to say? Suppose that we do postulate a Cause (dignifying this, like its putative effect on the universe, with an initial capital). Then, since it is, by the hypothesis, beyond the range of scientific investigation, the most that can be said is—emptily—that it must have been sufficient to produce its effect: "Not only the will of the supreme being may create matter", as Hume went on to say, "but, for aught we know *a priori*, the will of any other being might create it, or any other cause, that the most whimsical imagination can assign".

Here, as so often, Lewis and I come very close to one another. (I am reminded of numerous discussions with my longtime Keele colleague, Professor D. M. Mackay: our undergraduate audiences were always astonished and, we hoped, instructed to find us in agreement about almost everything except the central issue in debate!) Where I speak of "either an unknown cause or no cause at all", Lewis believes that we are required "to postulate some transcendent reality"; a reality which "by its very nature as we postulate it, can convey nothing to us . . . since everything we can understand must be in terms of finite existences. . . ."

May we say that for Lewis reason merely "requires us to postulate some transcendent reality"; but that, for any further knowledge of that transcendent reality, we have to refer to revelation in "holy scriptures"? If this is indeed his position, then it would, I think, be entirely consistent to maintain that it is only thanks to that revelation that he becomes able to identify what appears to him to be experience of God as in truth the genuine article.

There is no room here, nor is this the proper occasion for us to continue our long running disputation as to whether the subject of human consciousness "is a creature of all too solid flesh and blood" (Flew) or "a severely mental existent" (Lewis). In any case it is an issue which we have each of us addressed in series of Gifford Lectures, given and published. It is, nevertheless, very much to the point to suggest that to establish his contention would be to erect a massive obstacle to the advance of scientific naturalism. This was clearly understood by, for instance, Pope Pius XII when in the encyclical *Humani Generis* he wrote: ". . . the teaching of the Church leaves the doctrine of evolution an open question, as long as it confines its speculations to the development from other living matter of the human body"; nevertheless, "that souls are immediately created by God is a view which the Catholic faith imposes on us".

Comment on Lewis and Flew

• • • • • • • • • •

Professor Hugo Meynell

Professor of Religious Studies at the University of Calgary. Works include *God and the World* (1971) and *The Intelligible Universe* (1982).

• • • • • • • • • •

Professor Lewis tries to persuade us that some being other than the universe is needed to account for its existence; and that this being is at least something like what has always been known as God. Professor Flew is convinced that at best there is no need for us to believe in the existence of God; and that at worst the very supposition of the existence of such a being is confused.

It is rightly stressed by Lewis that what ultimately *explains* the universe of creatures must be very different in kind from anything *within* that universe. But unless there is at least a remote analogy between the nature of the alleged "creator" and at least some creature, it is difficult to see how we can significantly speak of a creator at all. The more seriously one takes the doctrine of divine transcendence or "otherness", the less prospect there appears to be even of making sense of talk of God in terms of earthly language, let alone of giving any earthly reason for believing that such a being really exists. Surely Flew is right at least to the extent that on this issue the ball is very definitely in the court of the theist. My own view, which is indeed a very ancient one, is that the closest analogy to God and to divine creativity within our experience is intelligent human agency. God is supposed to conceive all possible states of affairs and to will the universe we actually have, rather as we ourselves conceive and will our comparatively small-scale actions and products. Rather as you or I can think of a number of ways of writing the next sentence of the letter or article that we have on hand, and can write it in one of these ways or not write it all; so God can bring into being this universe, or another, or can fail to bring anything else into being at all.

The rational theist owes her opponents a clear explanation of how God is supposed to "transcend" the universe, to be beyond, behind, above it, or whatever. It is easy to say, that by these expressions one means simply that God is *other than* the universe as a whole or any part or aspect of it. But given what is at least a natural interpretation of the term "the universe", as "the sum total of all that there is", it is at first sight very difficult to see how what is other than the universe as a whole or any part or aspect of it can be other than nothing! (Flew has made this point with force and elegance in several of his writings.) It appears to me that someone who wishes, in the face of this challenge, to make sense after all of talk of God as both transcendent and existent, has to ask a question something like the following: Within the universe-A (the sum-total of what exists), is it possible or likely that there exists some entity which is related to the rest of what exists (universe-B) rather as a cause is related to its effects, or as a human agent is related to her actions and products? In that case God will, if God exists, *be a part of* universe-A, but "transcendent" of universe-B—that is , identical neither with it as a whole nor with any part or aspect of it.

Now it is clear enough that, unless one presupposes certain rather primitive conceptions of God, God is not supposed to be an *embodied* agent. But it may well be objected that a non-embodied agent is indistinguishable from no agent at all. This issue is a complicated and ramifying one into which there is no space to go in detail here. But it seems to me that there are conceivable cases (some actual ones have been reported, perhaps always mistakenly or mendaciously) in which to postulate a non-embodied agent would be the reasonable thing to do. Suppose my pen were on occasion firmly to detach itself from my fingers while I was writing, a piece of paper to place itself on my desk, and the pen to write in a consistent scrawl such comments as, "The paragraph which you have just composed is philosophically incompetent". I think that the activity of a non-embodied agent would be about the most sensible way of explaining such a series of events. (It may be noted in this connection that the postulation of what we cannot observe to explain what we can observe is a standard procedure both in science and in ordinary human affairs; one thinks of fundamental particles in physics, of events in the remote past, and of the thoughts and feelings of persons other that oneself.)

But however this may be, I think that our knowledge and awareness of ourselves as rational agents is instructively different from our knowledge and awareness of ourselves as physical organisms. This should give some pause to those who assume as a matter of course that to be a rational agent is necessarily also to be a physical organism. Flew acutely remarks that to acknowledge the actual or possible existence of unembodied souls is to set limits to scientific naturalism. Now it appears to me that "scientific naturalism" itself is a complex of ideas and convictions not without ambiguities and anomalies and that these have a good deal to do with the question of whether a rational defense of theism is possible. To bring out this point, I would like to distinguish between two possible versions of "scientific naturalism", one of which makes the operations of human reason itself a reflex of physical and chemical events on the brain, the other of which admits reason to be autonomous. The first kind of scientific naturalism is certainly more often implied than it is clearly stated, as it was by the zealous follower of Freud who declared that his master had shown that the human mind was no better adapted to finding out the truth about things than a pig's snout. When this view is distinctly set out, its self-destructive consequences become obvious. One should presumably only hold to scientific naturalism because it is reasonable for one to do so; so it would be perplexing indeed if one of the implications of scientific naturalism was that *nobody* maintained the truth of this account of things, or of any other, because it was the reasonable thing to do. To accept scientific method as liable to lead to the truth about the universe, in fact, is to be committed to the view that reason is sufficiently autonomous, and to that extent independent of its material basis in the human brain, to think thoughts and to make statements because it is reasonable to think that they are true.

Let us then consider the second kind of scientific naturalism which we have distinguished, which affirms the autonomy of human reason. Now it is remarkable that science (at least as usually understood) presupposes not only that human reason is thus autonomous, but that the universe to its farthest reaches in time and space is permeable to such autonomous reason. (I am not taking into account those views of science which I consider to be degenerate and confused, that consider it as merely a matter of controlling our environment, rather than discovering the truth about it.) Humanity is apparently a tiny speck of protoplasm clinging to the surface of a small body circling a very ordinary star which is one of billions in a galaxy which is itself one of billions; and yet it has the conceit to think that the products of its miniscule brains can aspire to the truth on such matters as were at issue in the last half-sentence. As the Stoics declared long ago, the *logos* within the human mind finds itself confronted with a *logos* within the universe as it exists and has existed prior to and independently of the human mind. I have argued elsewhere that it is this permeability of the universe to human intelligence, rather than the mere existence of the universe, which provides the best grounds for thinking that there is *within* or *beneath* it (according to whether one is thinking in terms of universe-A or universe-B) something analogous to human intelligent will. The divine intelligence accounts for the intelligibility of the universe, the divine will for the fact that it has the particular intelligibility (in terms of oxygen rather than phlogiston, of relativity rather than a luminierous ether) which scientists progressively discover it to have. It looks as though the first type of scientific naturalism that we distinguished is self-destructive; and that the second leads beyond itself to the conception and affirmation of a being that is in some sense supernatural (in the sense of other than universe-B or any part or aspect of it). I would maintain that the intelligibility of the universe is the main reason for doubting the obvious inference from the fact pointed out by Flew—that *within* the universe-B we have rather good reason to suppose, in opposition to the "animism" characteristic of many human cultures, that explanations in terms of intelligent agents ought to be confined pretty strictly to the realm of human affairs.

Both Lewis and Flew allude to the following problem, the latter considering it fatal for the prospects of a rational theism, the former not. If the universe, or perhaps rather the fundamental laws and initial conditions underlying the (rest of the) universe, are supposed to be in need of explanation, why should not the same apply to God? If the universe, or these basic features of it, are objectionable as brute facts, why should God be any more acceptable? As Lewis admits, there is something profound as well as naive in the question of the child, when informed by her mother that God made the world, "But Mummy, who made God?" The implied objection is a perfectly reasonable one; but I believe that the rational theist can meet it. Suppose God is, as I have said, that on the understanding and will of which *all* else depends. In that case, God, in virtue of being God, *could not depend on* any other being or beings. The rational theist may thus claim that God is *required* in explanation of the otherwise "brute facts" of the world, without being such as to require explanation in turn. It is not that God, as envisaged in the way that I have suggested, could not conceivably not exist at all (as in so-called "ontological" arguments); but that, *if* God existed, God's existence could not be dependent on that of any other being.

In our immediate environment at least, the right explanation of an intelligible state of affairs (like the appearance of a rude remark written on my blackboard, or the arrival of a letter demanding my immediate attention) which *happens* to be the case, but does not *have* to be the case, and is otherwise impossible to explain, is intelligent agency. The rational theist maintains that this applies to the intelligible state of affairs which is the universe-B as a whole.

One Word More

· · · · · · · · · ·
Professor Antony Flew

Since, notoriously, all good things if not all things without exception must come to an end, I will make only three comments on Hugo Meynell's "Comment".

1. Meynell suggests "conceivable cases . . . in which to postulate a non-embodied agent would be the reasonable thing to do". Certainly the occurrence of such cases would reasonably lead us to explore the possibilities of constructing a concept of— to revive the traditional terminology—an incorporeal spiritual substance. But, before it could be conceded that the construction had been satisfactorily completed, it would have been necessary to specify how such an hypothesized cause of those conceivable phenomena was to be, at least in principle if not in practice, first identified as actually present and active and then subsequently reidentified as having remained one and the same spiritual substance through the intervening—as the lawyers say—effluxion of time.

2. Recognizing that "there is something profound as well as naive" in the child's question, "But Mummy, who made God?", Meynell proposes a definitional stop: "Suppose God is, as I have said, that on the understanding and the will of which *all* else depends. In that case God, in virtue of being God, *could not depend on* any other being or beings". Fair enough: Meynell and his lady friend, "the rational theist", are fully entitled to appeal to this element in their definition of the word "God". What is sauce for the goose, however, must necessarily and by the same token be sauce also for the gander. So it is equally open to the Stratonician atheist to pre-empt this definitional response by a previous definitional stop of his own. For he (or she) is no less entitled to insist that the existence of the universe (the theists' universe-B) with whatever are in fact found to be its most fundamental characteristics necessarily constitute the ultimates of explanation. We have thus a definitional preclusion of the question, "Who made the universe?" parallel to the proposed definitional preclusion of the question, "Who made God?"

3. It is , I think, important to bring out more clearly than Meynell himself does what grounds are his for maintaining that "the universe-B as a whole" constitutes an "intelligible state of affairs". For in arguing that "In our immediate environment at least, the right explanation of an intelligible state of affairs . . . which *happens* to be the case, but does not *have* to be the case, and is otherwise impossible to explain, is intelligent agency", Meynell's examples are of occurrences normally known never to occur without benefit of human agency. But in maintaining that "the universe-B as a whole" constitutes an "intelligible state of affairs", Meynell's assertion of intelligibility presumably excludes no more than total chaotic irregularity. His argument here therefore appears to reduce to a revised version of the traditional Argument to Design, the objections to which are all too familiar.

Another Word More

· · · · · · · · · · ·

Professor Hugo Meynell

1. It seems to me that we can easily imagine what it would be like to have reason to identify and re-identify what Professor Flew calls a "spiritual substance"; and I do not see why permanent ownership of the same parcel of matter, however palpable or rarefied, should necessarily have anything to do with it. (To say so much is to open a vast can of philosophical worms; I can here only state a conviction which needs defending at considerable length.)

2. I concede to the Stratonician atheist that we may properly mean by "the universe" "the sum total of what exists", and that if we mean this by "the universe", the question "what caused the universe?" makes no sense. But I do not see how she can then properly rule out the question, "Do we have reason to suppose that, *within* the universe as so conceived, there is that which is related to *the rest of* the universe as cause to effect, or as conscious agent to action or product?"

3. The argument from intelligibility seems to me quite distinct from the argument to design and to stand or fall independently of it. It applies, to take one instance, just as much to the ultimate matter out of which the universe may be deemed to consist, as to those forms or structures imposed on or evolving out of this which are the starting-point of the argument to design.

Concluding Comment

Professor H. D. Lewis

I think we are now at the stage where it is better for other contributors to come in. The main thing for me in Professor Meynell's observations is the difficulty he finds (and which others share) in making any intelligible statement about (or making any sense of) a reality alleged to be beyond anything we can speak of or understand in the terms we use to refer to normal finite reality. I have little to add to what I have said on this issue earlier. The ultimate incomprehensibility of God in himself, and the essential mystery of the way he works in itself, is stressed in many religious scriptures, including the Bible in many parts. That we feel impelled, by finite experience and the finite world, to recognize the necessity of such reality is not ruled out by the inevitable failure to say more about it in itself. To amplify this would be to repeat much that I have already said in my initial article. But the ultimate mystery for us of the being of God in himself does not require us to rest with this and affirm nothing further in other allusive ways. I have myself stressed the analogy of our awareness of one another, and this seems to me to take proper note of the place of "intelligent human agency" in our understanding of God and his ways, so far as this is possible for us. We must not make such analogies final but allow them to provide a genuine clue to the things we say about divine intervention and concern. How this proceeds is what I have considered all too briefly in the closing part of my paper, and I shall return to it as a central religious issue in itself elsewhere. I do so in part in my contribution to the volume in honor of Professor Ivor Leclerc. But I must say here than the inability to speak of God in himself in specific and final human terms, and with our own full understanding, is not an insurmountable difficulty for those who claim to have impressive specific knowledge of God in other ways than the insight, allegedly required, which would put us in some ways at the level of God. We are finite creatures, but I see nothing preposterous in the notion that God can communicate with us, at our level of

251

understanding, in vitally significant ways and build up our fellowship with him without our being able ourselves to rise beyond finite existence.

I wish I could venture to say more on this, but perhaps others will, and I undertake to return to this issue in due course elsewhere.

PART FOUR

· · · · · · · · · · ·

Concluding Scientific Postscripts

1
The Origin of the Universe in Science and Religion

Professor William Stoeger

- Born 5 October 1943

- Ph.D. in astrophysics, University of Cambridge, 1979

- Currently Visiting Davies Professor, University of San Francisco, and Staff Astronomer, Vatican Observatory

- Areas of specialization and accomplishments: astrophysics; cosmology; boundary conditions; works include *Philosophy in Science* (M. Heller, W. R. Stoeger, and J. Zycinski, eds., 1982); *Physics, Philosophy and Theology: A Common Quest for Understanding* (R. J. Russell, W. R. Stoeger, and G. V. Coyne, eds., 1988); and over forty papers in scholarly scientific journals

The issue of origins has always been a preoccupation of religion and theology, as they have attempted to provide us with the ultimate meaning of our existence and that of the world around us and a stance toward that world and towards our fellow human beings which enables us to live fruitfully in the present, while moving hopefully towards a destiny only dimly perceived. The sciences, in contrast, did not treat of origins—not until they began to reveal our world and our universe as evolutionary. They presupposed origins, presupposed existence, and investigated in great qualitative and quantitative detail the entities, structures, and processes which constitute the material world and the laws which govern them. With the advent of the evolutionary perspective, the natural sciences (first biology and geology, and later physics, astronomy, and cosmology) have quickly found themselves dealing with this fascinating, mysterious, and sometimes unsettling area, which was once the sole preserve of theology and philosophy. Until recently, biology, chemistry, and physics have dealt with origins precisely by describing and explaining how new structures and processes, of greater complexity and specialization, have arisen from less complex realities: how baryogenesis occurs from the quark sea, how hydrogen is transformed into heavier elements during primordial nucleosynthesis (about three minutes after the Big Bang in the standard cosmological model and in stars later on), how relatively simple molecules develop into self-replicating ones, how collections of self-replicating molecules eventually develop into amazing new entities like cells, and so on—moving from one already existing complex of realities governed by well-defined laws to another more complex set of realities governed by higher order laws, thus explaining their origin. Recently, however, in contemporary cosmology, as scientists have investigated the very early universe at ever higher energies—which correspond to times ever closer to that of the Big Bang—and the processes which dominate at those energies, there has been a growing sense that physics, cosmology, and relativistic field theory are all beginning to treat ultimate origins—the origin of space and time, of matter and energy, of the physical laws themselves. Certainly, that is very strongly suggested by the superstring program, which is often described as moving towards "a theory of everything", and by the complementary programs in quantum cosmology which search for a way of starting off the universe without having to specify initial conditions—such as the Hartle-Hawking ground state proposal, or similar alternatives.

So now, as physics and cosmology move in (seemingly) on ultimate origins, is there any room for theology and religion? Is there any room for a Creator, as Hawking, Sagan, and others have asked? In this brief paper I shall confront the approach to the origin of the universe as it seems to be emerging in contemporary physics and cosmology with the usual philosophical treatment of *creatio ex nihilo*, which provides the basis of the standard Christian theological approach to the problem of ultimate origins.[1] I shall also apply this, briefly, to other related issues, including the anthropic principle, continuous creation, and in a general way to other theories of the early universe, which I shall treat together. It will become apparent that the most basic sense of *creatio ex nihilo* does not conflict with scientific treatments of the origin of the universe, but is rather a way of dealing intelligibly with whatever they must presuppose—and they must all presuppose something.

Among the specific questions I shall address in relating *creatio ex nihilo* to the emerging scientific accounts of the beginning of the universe will be whether the Big Bang as it is characterized in cosmology can be identified with "the moment of creation" in philosophy and theology—the point at which absolutely nothing becomes something; whether a theory of everything within the natural sciences is really possible, from a philosophical point of view; whether God, if he/she is needed for the creation of the universe, had any choice in the way He/She created it, as Stephen Hawking asks[2], and as Einstein did before him; the problem of the origin of physical laws; whether the multiple universes scenario really alleviates the issues addressed by *creatio ex nihilo* and the anthropic principle, without appealing to a necessarily existing, primary cause (God). But first, before entering into these questions, it is important to clarify the terminology which is used in these scientific and philosophical contexts, particularly the meaning of the word "creation" and its temporal connotations.

. . . .

What is "Creation"?

In science the word "creation" is used rather loosely. Whenever a new particle appears as the result of an interaction of other particles or by polarization of the vacuum (vacuum fluctuations) we speak of its creation. In other contexts, we speak of stars being created out of collapsing and fragmenting clouds of hot nebular gas, of proteins being created out of amino acids according to the genetic blueprints carried by messenger RNA, of a zygote being created from the fusion of ovum and sperm. But, from a philosophical point of view, and as we shall use it here, we mean something much narrower and more specific by "creation". We mean *creatio ex nihilo*, both the *ultimate* bringing into, and maintaining in, existence—creation out of absolutely nothing. The creation with which we are familiar in the sciences—with which they deal and which they are capable of describing and explaining on the basis of the theories which they elaborate—is not *creatio ex nihilo*, but rather the transformation of previously existing material or physically accessible entities into qualitatively new ones. A photon, for example, becomes an electron and a positron through "pair creation". The positron and the electron are not necessarily existing entities—they do not of their nature have to exist, they do not explain their own existence. In each case their existence depends on something else, in this case the photon from which they were created, along with the whole causal chain upon which the photon depends. It becomes clear that every physical entity or object is not necessary but only contingent. No physical entity explains itself or its own existence. Each can be traced to prior or more fundamental ones. And those in turn to others. And so on *ad infinitum!* Thus, at some temporal or logical point we must explain the transition from nothing to something. Why is there something rather than absolutely nothing? And, if we answer that question, then why *this* rather than something else?

· · · · · · · · · · ·
The Basic Content of Creatio Ex Nihilo

Thus, from this description of creation we can see that the principal content of *creatio ex nihilo* is simply to underscore the insight that the existence of something, whether it be energy, material particles, or operative laws, requires a cause which either necessarily exists in itself, or ultimately rests on a cause which necessarily exists in itself—a primary cause, the first in the causal chain, not needing any other cause to explain its existence. To have a chain of contingent entities stretching back into the past, none of which explains its own existence and none of which causally depends on an entity or cause which explains its own existence, is simply unintelligible. Or to put it another way, to trace the underlying causes of an entity to more and more fundamental entities and processes which themselves depend on other entities, causes, and processes for their existence does not fully explain the existence of the entity in question, unless the search terminates in an entity or process which explains or causes itself, which *necessarily* exists, which of its nature cannot not exist. If it terminates, or seems to terminate, in an entity—a geometric manifold, a symmetry group, some "cosmic egg"— which is not necessary in this sense and we simply throw up our hands and say that this entity simply exists, simply *is*, without searching for any primary cause upon which to rest the "cosmic egg", then we have prematurely and arbitrarily abandoned our quest for the intelligibility of the whole chain.[3] Then we have no explanation whatsoever for the ultimate cause of the existence of anything—or for why *this* symmetry group or manifold exists, and not some other imaginable one, or why this particular one which we can mathematically investigate is given concretization within reality rather than some other one. If it is not necessary, as symmetry groups and geometric manifolds are not, then there is nothing in its nature which specifies that it must be concretized or instantiated in physical reality.

This is really the same question Stephen Hawking asks towards the end of his book: "What is it that breathes fire into the equations and makes a universe for them to describe? . . . Why does the universe go to all the bother of existing? Or does it need a creator, and, if so, does he have any other effect on the universe? And who created him?"[4] If the universe as we know it, or as we come to know it at some stage of its previous history (for example, in its "ground state", as suggested elsewhere by Hawking in the volume and in his previous work), explains or necessitates its own existence by its very nature, then this would be perfectly obvious, and not obscurely hidden behind veils and veils of contingency. Hawking clearly seems to recognize this, while not fully admitting it: leading to the above key questions he poses, which essentially demand a general philosophical answer like *creatio ex nihilo*, he comments: "Even if there is only one possible unified theory it is just a set of rules and equations. . . . The usual approach of science of constructing a mathematical model cannot answer the questions of why there should be a universe for the model to describe".[5] In other words, any physical model of the universe—even though it be the definitive unified field theory—will apparently not be able to account intelligibly for the existence of the universe by itself. It will need some underlying existential ground or support, which must be presupposed by science and which science itself, as it is presently conceived and practiced, cannot completely reveal. This existential ground or support is exactly what *creatio ex nihilo* purports to supply. It does not, of course, completely reveal the ground or support, but it does argue to the necessity of its existence, and in further philosophical analyses which I shall not go into here, to certain broad characteristics it must possess, such as simplicity (not being composed of parts), uniqueness, immateriality, perfection.

It should be clear, furthermore, that this primacy is fundamentally *logical* or *ontological* primacy, not necessarily *temporal* primacy, although the latter obviously implies the former. In other words, *creatio ex nihilo* does not necessarily imply or demand a beginning in time, nor a beginning of time. Such a beginning is necessary, but not a sufficient condition, for *creatio ex nihilo*.[6] The primary cause—God, if you will—can support in existence a temporally infinite (without beginning or without end) universe, all of which is causally contingent.[7] Christian revelation, as it is contained in the Bible and in Church Tradition, is usually interpreted as indicating that there was a temporal beginning.[8] But this is *not* philosophically essential, and is not part of the core of the doctrine of *creatio ex nihilo*. The act of bringing something out of absolutely nothing did not have to happen at a given temporal point, or points. Neither time nor the material universe had to have a beginning. It is important to notice, in this context, of course, that if the universe has a beginning in time—or if time itself has a beginning—that this would be a clear indication, sign, or symbol of *creatio ex nihilo*, a physical manifestation of it.[9] If the universe is temporally infinite in both directions, *creatio ex nihilo*, while still very much operative, would lack a clear manifestation in a special event, even though every aspect or constituent of the universe would clearly manifest its contingency by depending on others for its existence, either logically or logically and temporally, by not existing necessarily.

But the argument is also very incisive in the other direction. The act of bringing something out of nothing (as long as the something is contingent) is needed at every temporal and spatial point. The primary cause cannot cease to act in support of secondary, contingent causes, processes, or entities—cannot cease to be influential and effective—without the universe, which is contingent, slipping back into absolute nothingness. In fact, the very continued existence of physical reality is an incontrovertible sign of the continued action, the continued effectiveness, of the primary, necessary cause. This aspect of *creatio ex nihilo* is often described as "God's conservation of creation in existence". From this perspective, even if there is a beginning of time, that "point" is fundamentally no more special than any other point. The problem of the origin of the universe, philosophically speaking, is not basically that of a temporal origin, but rather that of a "causal origin" which renders, first of all, its existence and, secondly, its characteristic order intelligible—which adequately explains or accounts for both.

The care and precision I have used (within the severe limitations imposed by the awkwardness of any language in speaking of things like the ground of existence) must also be applied to two other concepts employed in describing *creatio ex nihilo*, namely "nothing" or "nothingness" and "cause." I have used the phrase "absolute nothingness" to emphasize the fact that the *nihilo* to which *creatio ex nihilo* refers is absolute nothingness, not just the vacuum of physics. There can be absolutely nothing—no matter, no energy, no laws, no concepts, no mind, no time, no space, no manifolds, nothing—no laws of statistics, probability, quantum fluctuations, no context of logic or order, except—and here is where language has already broken down—a necessary being, a primary cause. To put it more correctly, besides the possible existence of a necessary being there is absolutely nothing else— no contingent beings. Now, in the case where there is not a necessary being, then absolute nothingness will remain absolute nothingness "forever". There is no intelligible way in which this nothingness can yield something—existence, something existing. In the case, the radically different situation, in which there is a necessary being and absolutely nothing else there can be something else which is not the necessary entity caused or brought into being by the necessary entity. Thus, working the argument in the other direction as we have already done, we can say that the existence of any contingent, non-necessary entities implies the existence of a necessary one, a primary cause, to endow other things with existence—to breathe fire into them, to discover in them something worthy of existence, and to give them that existence, to instantiate or concretize them.

When we come to the word and concept "cause", analogous qualifications have to be applied when we use it for the necessary entity, the primary cause. The causation exercised by the primary cause is of a completely different order than that exercised by the contingent, secondary causes with which we normally deal, and which are the domain of the sciences. If anything at all exists, then the primary cause must be acting, endowing everything else with existence, enabling contingent causes to transfer existence to other entities along with their specific qualities. But the action of the primary cause, as primary cause, is so pervasive and continuous or at such a fundamental and radical level that it cannot be individuated or isolated from other causes in the way in which contingent causes are. Without its activity, the contingent causes would not exist. Absolutely nothing would. So it provides the overall, subtle causal and existential context which is essential for everything else. We normally take it for granted— the sciences, as I have already mentioned, presuppose it—and rarely notice it, just like the air we breathe. But, without it, everything would cease to exist.

This principle *creatio ex nihilo*, which I have just elaborated in its fundamental import, is, it should be pointed out, a philosophical conclusion, the result of philosophical analysis of reality as we know it, and not a religious or theological principle. It has been, however, appropriated by Christian theology in order to articulate and render intelligible what is believed to have been revealed by God in Scripture and in Tradition concerning creation, the contingency of all that is not God, and his or her irreducible and critical role in holding creation in existence. The content of *creatio ex nihilo* itself, though so very important, is, as has been frequently pointed out, extremely bare and unadorned, once it has been unveiled.[10] There are no details, no motivations, no temporal or spatial specifications, not even whether or not there was a beginning of time—just the bald insight that a necessary and self-sufficient entity or cause must ground anything that exists, whether it has a temporal beginning or not. But then such baldness and abstraction is characteristic of philosophy in general, and of metaphysics in particular. It is left to the sciences, to the arts, and to other areas of human imagination and endeavor to fill in the details on all the different levels of being and action.

· · · · · · · · · ·

The Relation of Creatio Ex Nihilo to Contemporary Cosmology

Now I wish to go on and reflect on how *creatio ex nihilo* as I have described it impacts the findings, the solid and provisional conclusions, and the speculations of contemporary cosmology. I thoroughly realize that many may not accept the justification I have given for it—which is pretty much the argumentation which is usually given—on philosophical, linguistic, or other grounds. But granted that this is what it means, and granted for the sake of this presentation that it is basically a valid principle, then how does it critique emerging cosmological thinking concerning the origin of the universe? There are a number of aspects of cosmology, a number of very general ones and a few important particular ones, which can be fruitfully discussed in reference to it.

The first general comment flows from the baldness and unadorned character of *creatio ex nihilo*, which we were just discussing. *Creatio ex nihilo* is not in competition with physics and cosmology, as they are presently understood and practiced, in describing the origin of the universe and its evolution to its present configuration. Rather it complements what physics and cosmology discover by justifying—or rather, giving a general, rather heuristic grounding of—a presupposition in the scientific treatment. Cosmology will find itself in conflict with *creatio ex nihilo* only if it succeeds in developing a theory, a "theory of everything" in the strict sense, which specifies its own concretization and instantiation in reality as necessary, self-sufficient and unique—which of itself "breathes fire into the equations and makes a universe for them to describe". But this seems very unlikely. Many researchers in the field, and Hawking himself, seem to admit that this is beyond the realm of possibility. If, however, it turns out that such a theory can be constructed and is correct, then it would embrace not only science but the contents of philosophy itself, incorporating even more than the bare *creatio ex nihilo* contains—for it would contain also a detailed characterization of the primary cause which endows all else with existence as well as a complete description of the operation of that primary cause, of its essential activity and relationships with respect to all contingent entities and processes.

Assuming that this is not possible and that the present relationship of *creatio ex nihilo* to contemporary cosmology continues to obtain, we further see that it is consistent and coherent with any scientific theory of the origin of the universe which does not pretend to be a "theory of everything" in the radical sense as we just described it. *Creatio ex nihilo* says nothing about the processes, laws, succession of evolutionary states, the essential features of space and time, quantum indeterminacy and fluctuations with which very early universe cosmology is concerned. In fact, as we have also seen, it does not even specify whether or not the universe (and time itself) have a temporal beginning. A universe which always existed and one which has a beginning in time are both consonant with *creatio ex nihilo*. And so, for example, it is also consonant with Hawking and Hartle's ground state manifold having existed forever and then suddenly blossoming into space-time as we now know it. What *creatio ex nihilo* demands, however, is that there be a primary cause always giving that ground state manifold existence and existentially supporting the mathematical and physical laws it obeys, as well as the laws—whatever they would be—describing its blossoming into the Big Bang. It seems clear that such a primordial, "timeless" manifold would not contain the principle of its own existence. The affirmation that it does is simply not an intelligible statement, given what we know of manifolds.

Thus, *creatio ex nihilo* is consonant with continuous creation, the inflationary universe scenarios, all of the various unification schemes, including superstring theory, quantum fluctuation theories, the plasma physics approach of Hans Alfven and his group, and, as I have already said explicitly, the ground state proposal of Hawking and Hartle. To put it negatively, *creatio ex nihilo* is a much too abstract and large-mesh philosophical principle to bring any critical precision to bear on the scientific and even the finer-mesh philosophical details and issues involved in the various competing cosmological scenarios for the origin of the universe, as it is generally understood within physics and cosmology. And thus it is incapable of providing criteria for distinguishing among them—among the suggestions for the detailed processes and structures which may have been involved.

It should be pointed out here again in this context, of course, that physics and cosmology, when they do deal with the origin of the universe, rarely do so directly, and always in terms of the origin of certain pervasive features of the universe, like space and time, or space-time, matter, or rather mass-energy, and so on. And in investigating their origin they must inevitably do so in terms of some pre-existing entities or structures or some pre-existing set of laws which these obey. Even the vacuum in physics is such a structure and obeys certain laws. It is not absolutely nothing. Neither is geometry—nor the principles of mathematics and logic. The origins with which science can deal are always what we might call "relative origins", which are indeed very important, absolutely crucial, for us to understand. But they are not absolute, or ultimate, origins. This is because, as I have already pointed out, the natural sciences must always presuppose the existence of something to study and an order or regularity which characterizes the behavior of that something, such that it can be described by a body of laws—such as the laws of conservation of mass-energy, or of momentum. If something exists but is absolutely chaotic and unpredictable in its behavior, science would be at a loss from the start: it would be impossible. So the sciences as such are incapable—at least as their methodology now is—of describing or characterizing absolute origins, or of justifying or grounding the assumption of existence and the assumption of order. They presuppose that something exists and is orderly, and they then discover that these two presuppositions are supported by what is revealed or discovered in their investigations on the basis of these two assumptions. They are not disappointed in their quest—in fact, they are strongly reaffirmed in the way they have proceeded and in the assumptions they have made. But at the same time science is unable to determine *why* what exists is regular and orderly. The ultimate sources of existence and regularity are inaccessible to scientific techniques and methods. That is precisely why "what is the origin of physical laws?" is such a difficult, and very important, question on the boundary between physics and philosophy.

It is clear why the sciences cannot deal with ultimate origins and cannot determine why what exists exists and why it is orderly and not completely chaotic. In order to do so, science would have to detect, discern, reveal, and describe absolute nothingness or absolute disorder as well as the entity or cause which moves "reality" from these states to the states of existence and order. This would be "the creation event", whether it was a temporally unique event, or a metaphysical "event" which is always occurring. Cosmology and the other sciences cannot disclose such an "event" simply because they are incapable of transcending the barrier between absolute nothingness and something, between absolute chaos and order: they are not methodologically established to have "one foot on each side of that divide". They do not have access to what is "before" existence and "before" order, simply because they have presupposed both basic existence and basic order and have not questioned the ground of either.

Did God as the primary cause have a choice in creating the universe as he/she did? This is one of the other questions which arise in this context. From what we have already seen, it seems clear that he/she must have had a choice. Lack of a choice follows only from a theory which dictates its own concretization (a theory of everything) which at this juncture seems very unlikely for the reasons given above, or from a theory which, while not dictating its own concretization, can be shown to be the only one which is able to be consistently concretized by the primary cause, for reasons of internal coherence. This also seems impossible to affirm. There are many different theories with adequate internal coherence.

The usual focus of speculation concerning the origins of the universe is the Big Bang. It is really what we might call the "scientific origin" of the cosmos—or rather the origin of the universe as it appears in the usual cosmological models. In these models, as we go back in time, we encounter a succession of hotter, denser phases. The limit, or "beginning", of these phases, which is given by the evolution equations—for example in the Friedmann-

Robertson-Walker (FRW) models, which are isotropic and spatially homogeneous—we call "the Big Bang". For various reasons I shall not go into here, it could not have been a single event. Rather, it had to be a whole manifold of "initial" events. Furthermore, cosmology and physics can really say very little about the Big Bang itself, because as it appears in most models—and certainly in the standard (FRW) ones—it is a boundary of the model, as well as a singularity, a place where important physical parameters become infinite—where the model breaks down. The Big Bang is thus, rather, the inaccessible limit of these hotter, denser phases, or time slices, of the universe.

In this context it is important to realize that the theory of gravitation which describes the dynamics of the universe, and therefore the evolution of the cosmology models, itself breaks down at times earlier than the Planck time, 10^{-44} seconds after the Big Bang. At this very, very early time, which corresponds to a very, very high temperature, conditions are such that a quantum theory of gravity, and therefore of space-time, must be used, instead of classical general relativity. As yet we have no such theory, although there are promising candidates under development, such as superstring theory. When we do, it will have to be used to treat those very first phases of the universe "right after the Big Bang", whatever they may have been—and whatever time, space, and event may mean, if anything, in that new quantum-gravity context.

It should be fairly clear from these considerations that the Big Bang as revealed by cosmology is not necessarily an absolute beginning—nor necessarily an absolute "beginning" of time itself. It is a "scientific beginning" relative to the models we now have, but probably will not be so according to the models which will rely on an adequate quantum theory. We cannot, of course, rule out that, prescinding from the consequences of such important improvements, this "scientific origin" may also be an absolute beginning—a point before which absolutely nothing existed, except the primary cause. But, even if it were, cosmology and physics as we now know them could never reveal it as such. In order to do so, as I have already pointed out, they would have to have some purchase on the state of being *before* the Big Bang, and to show that it was a state of *absolutely* nothing—no space, no time, no laws, no symmetry group, no primordial manifold. A scientific discipline is incapable of doing that. Science, and perhaps human knowledge in general, is incapable of disclosing that absolutely nothing prevails at a certain instant, or that such a state is "before" time began. Thus, cosmology cannot discern "the very first moment of creation", *if* there was one.

In fact, as we have already seen, there may not have been a very first moment of creation. Philosophical analysis does not require one. And it is not by any means essential to the *creatio ex nihilo* principle that there be one, though many standard interpretations of the Bible and Christian tradition point to some sort of a radical creation event (whether it really has to be an absolute beginning in the sense I mean it here is not at all clear). Certainly, if there indeed was such an absolute beginning in time, or beginning of time, before which there was absolutely nothing except God, then it would *imply* creation *ex nihilo*. But the implication does not go in the other direction. The Big Bang, however, is the only candidate we have for such a beginning, and therefore it can "symbolize" for us the beginning of the universe and *creatio ex nihilo* (since it is consonant with it). If it is used as such in contemporary myths of creation, fine. But it should be made clear that it *cannot* be identified with a beginning of creation or a beginning of time.

Is there any connection between *creatio ex nihilo* and the anthropic principle? Certainly, the primary cause, which brings something out of absolutely nothing, either from all eternity or at a single "event", can determine the ordering, character, and direction of what he/she creates, either directly or indirectly through the laws and constants he/she establishes to govern the evolution of that creation—and through the initial conditions he/she sets, if that is necessary. The primary cause *can* fine-tune the universe. But this is not a *necessary* consequence of *creatio ex nihilo* itself. It would be something in addition to it, specifying the kind of universe created, the range of possible entities which could arise within it, and embodying an overarching or long range teleology or finality on those laws and initial conditions. Even in a completely chaotic universe, without physical laws, the principle of *creatio ex nihilo* would be operative. Its *existence* would have to be adequately and intelligibly accounted for. *Creatio ex nihilo* is directed towards accounting for *existence*, not towards accounting for why what exists is the way it is.

Of course, confronted with an ordered evolving universe—in which amazing complexity on many different levels gradually emerges, in which living organisms come into being, in which conscious, knowing, socially oriented beings develop—one must search also for an adequate and intelligible explanation of such order and such rich potentiality. How is it embodied in the complexes of lower level laws in such a way as to give the appearance of overall directedness towards higher level entities? Can this be done without a long range teleology which, so to speak, seems to orchestrate evolution from outside the evolution and operation of the natural laws themselves? One would hope so. But, even if this can be demonstrated, the fundamental question remains: What intelligibly accounts for the embodying of such richly ordered potentiality in the complex of the lower order laws and entities in the first place? It seems that one could go beyond simple *creatio ex nihilo* and say that such ordering and apparent finality *does* require a basis in some primary cause which not only brings into existence but also endows with order and potentiality. This is just an application of the philosophical principle that effects (for example, an ordered evolving universe with great potential for complexity and for bearing entities capable of highly specialized behavior) demand an adequate cause, one which is capable of explaining what is observed. Thus, we could, as some have suggested, appeal to what we might call an expanded principle of *creatio ex nihilo* to explain or account for ultimately *both* the existence and the richly ordered potentiality of the universe. In doing this we would not be infringing upon the competencies of the sciences—we would be attempting to say nothing about any of the details of how it all unfolds, only insisting that its existence, order, regularity, and potentialities be grounded in an adequate cause, in an adequate ultimate explanation, however general and unspecified that may have to be. In saying this, I am speaking as a philosopher, not as a theologian or a believer.

The usual alternative given to the operation of a primary cause in explaining the fine-tuning of the universe and its bearing of life and intelligent life ("the anthropic principle") is some variation of the multiple universes scenario. But, in light of our extensive discussions above, is this really an alternative? I rather like the multiple universes scenario—in fact, I strongly suspect that it may be correct, though the sciences will probably never be in a position to verify it by experiment or observation (if there really are multiple universes, it seems very difficult to see how we would ever find out about the others observationally). But having multiple universes still does not begin to deal with the ultimate questions of why they exist to begin with and why as a collection or ensemble they have the order and potentiality they have for at least one of them to become a universe like ours. It merely postpones these ultimate questions to a previous step. An ensemble of universes by themselves cannot ultimately account for what exists or for the character of what exists.

In this brief essay I have attempted to present as clearly as possible the fundamental idea of the principle of *creatio ex nihilo* and relate it to some of the fundamental boundary questions which are arising in cosmology and physics today concerning ultimate origins. As is evident, what is claimed in *creatio ex nihilo* is very minimal, even in its expanded version, but very important. There are few details, just the insistence upon an adequate ultimate cause and the insight that such a cause cannot itself be contingent. Furthermore, such a cause may act from all eternity: there can be eternally existing contingent entities, an eternally existing universe. A beginning of time and a beginning of the universe is also possible, of course, and would perhaps be a clearer sign of both contingency and *creatio ex nihilo*—something definite we could point to—but it is by no means necessary. Finally, as I have already hinted, if the sciences are eventually to include an adequate treatment of ultimate origins, it seems that they will have to appropriate principles equivalent to *creatio ex nihilo* along with methods of analysis similar to those used in arriving at it.

· · · · · · · · · ·

Notes

1. Some of the points I develop in this essay have been previously discussed in my article, "What Does Science Say About Creation?" *The Month*, CCXLIX, No. 1148–49 (Aug./Sept. 1988), pp. 805–11. C. J. Isham, in his essay, "Creation of the Universe a Quantum Process", in *Physics, Philosophy and Theology: A Common Quest for Understanding*, ed. R. J. Russell, W. R. Stoeger, S. J., and G. V. Coyne, S. J. (Vatican Observatory, 1988 [hereafter *PPT*], pp. 375–408), has recently dealt with this topic in a very complete and penetrating way, describing the cosmological models of creation much more thoroughly than I am able to do here. My focus is intentionally on the philosophical side of the dialogue, as that is the one less adequately discussed in the usual treatments. Willem B. Drees, in his doctoral dissertation at the University of Groningen, *Beyond the Big Bang: Quantum Cosmologies and God* (La Salle, IL: Open Court, 1990), has discussed the most important issues in this confrontation with scholarly thoroughness and depth. See also his article, "Beyond the Limitations of Big Bang Theory: Cosmology and Theological Reflections", in *CTNS Bulletin*,

Vol. 8, No. 1 (Winter 1988), pp. 1–15. The conclusions I reach here are in substantial agreement with those of Isham and Drees, though my emphasis is somewhat different.

2. Stephen W. Hawking, *A Brief History of Time* (New York: Bantam Books, 1988).

3. Cf. W. Norris Clarke, S. J., "Is a Natural Theology Still Possible Today?" in *PPT*, pp. 103–123.

4. Hawking, op. cit., p. 175.

5. Ibid.

6. It is true, as Drees points out (see his dissertation, reference 1, Sections 2.5 and 3.4, and his *CTNS Bulletin* article, Section 3.2), that traditionally *creatio ex nihilo* has both a cosmogonic pole, pertaining to the universe's coming into being, and an ultimate dependence, primary cause pole. But the latter is certainly the more fundamental, as has been emphasized again and again, particularly in recent discussions.

7. St. Thomas Aquinas, *Summa Contra Gentiles*, I, 44. See also E. McMullin, "How Should Cosmology Relate to Theology?", in *The Sciences and Theology in the Twentieth Century*, ed. Arthur Peacocke (Notre Dame: University of Notre Dame Press, 1981), p. 39.

8. Ibid.

9. T. Peters, "On Creating the Cosmos", in *PPT*, pp. 273–96.

10. I am grateful to Gerald O'Collins, S. J., for emphasizing this point at a recent Vatican Observatory meeting on this subject.

2

Relativity, Quantum Theory, and the Mystery of Life

· · · · · · · · · ·

Professor Eugene Wigner

I have a deep interest in the basic problems of physics; this is, in fact, usual for physicists of our age. We realize that the existence of science is a miracle, a very new miracle if we compare its age with that of mankind, which is, we believe, not far from a million years. Physics, at least our day's physics, is less than four hundred years old if we consider its start with Galileo, Kepler, and Newton. But it has developed in the past four hundred years unbelievably fast, its area extended miraculously, and in fact it changed man's life almost fundamentally. This is a miracle, and we do not really understand why it started just four hundred years ago. I will not go into this question further, because I discussed it in some detail a rather short time ago.

Let me come instead to the subject which I do wish to discuss and which is quite important for the status of present-day physics. It is an inner contradiction in present-day physical theory, one part of which describes principally macroscopic phenomena and can be based on the general theory of relativity, the other part deals with microscopic systems and is based on quantum theory. The phenomena to the description of which these theories are usually applied are so different in shape and magnitude that the effects described by the other can clearly be neglected. Hence, the difference of their basic principles are rarely, if ever, perturbing in detailed applications. But I feel that these should be recognized since in a truly beautiful physics the two should appear as limiting cases of a more general and more profound theory. And, of course, there are sizes, intermediate, to which neither of the two theories can be applied. But these are outside the present interest of our science altogether.

I will first try to describe the limitations of the validity of general relativity when applied to non-macroscopic phenomena.

· · · · · · · · · · ·
The Microscopic Problem of General Relativity

General Relativity's basic quantity is the space-time distance of close-by points on space-time, described by the metric tensor, its components usually denoted by g_{ik}. The determination of these on the very macroscopic level is often discussed, but we want to consider its microscopic meaning.

Space-time points are, on the microscopic scale, phenomenologically defined only as crossing points of world lines. Such crossings are, of course, collisions, which influence the paths of the two colliding objects. But when we define General Relativity's metric tensor, we assume that one of the colliding objects is very large—perhaps a planet—the other very small—perhaps a light ray—so that the motion of the first one is not really influenced by the collision: its natural path can be determined. The situation is clearly different on the microscopic level, not only because of the difficulty just indicated, but much further because the point of collision is not determined on the microscopic level. The quantum mechanical description of the collision, the collision matrix, leaves the space-time point of the collision uncertain at least to the amount h/mc – m being of the order of the mass of the colliding particles—even though the collision matrix itself can be accurately measured—it is a "law of nature", not an "initial condition". Heisenberg suggested already in 1943 that it be considered as a fundamental concept of our physics, and F. Goldrich and the present writer showed (in 1972) that it can be determined, by repeated experiments, with arbitrary accuracy (assuming that the laws of nature have the usually assumed invariances, particularly invariance with respect to time and space displacement). This is fine as far as the collision matrix is concerned, which does not use the concept of space-time points, but it implicitly questions the usual formulation of the laws of nature which involve the possibility of the observability of quantities at space-time points. This can well be questioned. It is possible, and perhaps even likely therefore, that the usual geometry of space-time, the measurability of quantities at space-time points, will be abolished just as present-day quantum mechanics abolished the point-like structure of phase space. This possibility is supported also by the rather old (1949) observation of T. D. Newton and the present writer that even present quantum mechanics denies the existence of localized states for most elementary particles. If a particle were, at a given time, surely at a definite point of space, this would be true, if viewed from any coordinate system obtained by a Lorentz transformation around the space-time point specified. It has been proved in the aforementioned article that no such state exists for elementary particles—in fact, in order to approach the wave function of such a state, it must be the superposition of many wave functions representing different states of the system considered. This, a fact of purely quantum mechanical origin, also indicates that our usual space-time concepts will have to be basically modified.

The preceding discussion indicates that the basic concepts of the general relativity theory, in fact of common thinking, may have to be modified. We should not forget in this connection that quantum mechanics already modified our natural description of the state of a particle: this cannot be specified by giving its position and velocity. What is suggested in the preceding is that a further modification of that description is needed which abandons not only the naturally expected structure of the phase space but introduces a less definite structure also for its space-time components.

· · · · · · · · · ·
A Macroscopic Problem of General Relativity

Perhaps I should mention at this point also another problem of the general relativity theory, a problem which lies in the domain opposite to that just discussed; it concerns not the very small but the opposite, the almost infinitely large domain. The question is: What happened before the "Big Bang"? What originated it? After all, the time coordinate of general relativity should have no absolute limit; the area of the time coordinate should extend to infinity. Actually, this is a question which is, and was, present in the minds of all who were interested in the implications of the general relativity theory; it is a question which was not much asked because it was clear that there is no simple answer.

But a possible, and in my opinion attractive, answer was given not long ago by Paul Dirac. He pointed to the surprisingly large numbers which appear in our physics. One such number is the ratio of the electric and gravitational attractions between a proton and an electron; another is the age of the universe if measured in terms of a natural constant such as h/mc^2. Dirac postulated that these constants grew together and introduced new units for their specification. In terms of these units the Big

Bang occurred an infinitely long time ago, at $t = -$ infinity, so that nothing happened before. It is very possible that all of you are familiar with this idea of Dirac's, but I wanted to mention it because it may show that some of the fundamental difficulties of our physics may acquire solutions. We can hope for that (even though it would be terrible if all were solved). I now go to my last subject which, I must admit, I already discussed at least in part quite recently.

· · · · · · · · · ·

Quantum Mechanical Description of Macroscopic Objects—the Limitations

The fundamental problem of the General Theory of Relativity when extended to the microscopic level is, as was discussed before, the impossibility of the experimental determination of its basic concept, of the metric tensor. Quantum mechanics, as was implied before, also has a fundamental problem, but this enters if it is applied to the determination of the time development of a macroscopic system—the time development of the microscopic state thereof, that is, of its wave function. The difficulty is, fundamentally, that a macroscopic body cannot be kept isolated from the environment thereof, and the influence of the observer very soon comes in. This means that the microscopic state of a macroscopic body is essentially meaningless: it does not obey a causal law. This was, I believe, first pointed out by H. D. Zeh in 1970. I have made a calculation on the state of a cubic centimeter of W (tungsten) which I considered to be placed into intergalactic space, far from all matter. But even then, and even though its state is as little subject to outside influences as that of any other matter of similar size that I could think of, its microscopic state will be changed by cosmic radiation in about a millisecond. This means, essentially, that the time dependence of a macroscopic object's microscopic state is indeterminable, it cannot be fully described by quantum mechanics or any theory. Fundamentally, the validity of the causality, or of determinism, is limited.

From a fundamental point of view this is an important fact but, actually, we do not want to know the wave function of a macroscopic object. It would be so complicated, would have so many variables, that we could not specify it. This is worth realizing. We have approximate state vectors for many of them, and these describe, quite satisfactorily, some of the macroscopic properties of the macroscopic objects; and it is quite reasonable to assume that the macroscopic properties are the same for practically all state vectors of states with very closely equal energies. The macroscopic properties which have been calculated for several macroscopic bodies by such approximate state vectors include heats of vaporization of some solids and liquids, their structure and elastic properties. And many, many other macroscopic properties can be, and a good many have been, explained by our quantum theory. It is, in my opinion, nevertheless important to realize that a true microscopic description of macroscopic bodies is not possible now—and perhaps never will be possible. What can comfort us is that perhaps we will never want it: the truly detailed microscopic structure of macroscopic bodies may remain of no real interest. But, from the point of view of a philosopher, it is an important fact that Newton's initial conditions plus laws of nature duality does not have the same basic validity in microscopic physics as the earlier physics postulated. But it remains, practically, equally useful.

. . . .

Initial Conditions and Laws of Nature versus Quantum Mechanics

The concepts of initial conditions and of laws of nature formed a wonderful basis of Newton's mechanics. And, as was mentioned before, the classical description of the initial conditions can be relatively easily determined for macroscopic systems and so can their later states. This made it possible to verify their laws of motion. The general verification of the quantum mechanical time development equations for microscopic systems is much more difficult. It can be done only for states which we can create and the end state of which is verifiable.

Both processes, the production of a state with a definite state vector and the verification of the correctness of the state vector which the time dependence equation postulates the system to assume after a certain time, are supposed to be done by what we call the "measurement process". This determines, actually, the state which the system has assumed as a result of the measurement process (this will be discussed further in the next section). It follows that the verification of the time dependence equation requires two such measurements. The first one gives the initial state of the system, the second should be so chosen that it gives a definite value if the system has assumed, at the time of the second measurement, the state which the time dependence equation gives. It follows that the verification of the time dependence equation needs a repetition of the experiment—the second measurement can give the expected result with a certain probability even if the state vector at the time of the second measurement is not exactly the one given by the time dependence equation.

It appears that the full and direct verification of the time dependence of the state vector as given by quantum mechanics (originally by Schrödinger's "second equation") is very difficult. Actually most verification of the quantum mechanical theory comes from the determination of the "energy levels and transition probabilities" as initially postulated by Heisenberg, but much later (in 1926), as the basic concept. Perhaps I may mention that although an equation giving the energy levels and transition probabilities was first formulated by M. Born and P. Jordan, with the later cooperation of Heisenberg, the real formulation of the theory of the collision matrix is due to J. A. Wheeler (1937). Naturally, both the energy levels and the transition probabilities, and also the collision matrix, are implicitly given by the time dependence equation. But the opposite is not (the total time dependence of the state vector cannot be derived from them); it is not impossible to believe that it is not a fully valid concept. We can hope it will be proved to be one.

· · · · · · · · · ·

Problems of the Measurement Process

The quantum mechanical theory of measurement, so needed to verify the whole theory, has at least three weaknesses. At least two of these are generally admitted, though not emphasized. The first of these was mentioned already in the preceding section—it is the impossibility of determining the state vector, or wave function, of any system. This is not surprising and probably cannot be changed at all: the determination of the macroscopic state of a macroscopic system is possible only because the interaction with the measuring apparatus can be very weak, and since the apparatus can be practically microscopic, its state can be fundamentally influenced by a macroscopic system without significantly influencing the latter. This is not so for the microscopic systems.

The second weakness is that even if we believe we know the state of the microscopic system, it is in general not possible to verify it. The clearest state the verification of which is possible is the state of a single particle with a definite momentum. This can be verified by reflection by a lattice of known periods. But this is an exception. It can be shown in fact quite generally that most quantities are not exactly measurable, in particular not if their operator does not commute with all additive conserved quantities such as momentum, or energy, or angular momentum components, and perhaps some others. (This was demonstrated by me in 1952.) It is true, on the other hand, that this restriction applies only if an accurate measurement is postulated. But in many cases even an approximate measurement is impossible; the most striking case for this is the determination of the position of a particle with an accuracy which is natural to require, an accuracy which is a very small fraction of h/mc, where m is the mass of the particle.

The third weakness is due to the fact that measurements take time; as a result it is hardly possible to determine the state at the definite instant. As to the instantaneous nature of the measurement, this is made impossible already by the fact that practically all wave functions are extended in space, and the signals from removed points take time to enter. In fact, for this reason, and to confirm with the theory of relativity, it would be more reasonable to define the state of a system by its wave function on a negative light cone and the state after the measurement, on the positive light cone. The difficulty to even define the position of a particle at a definite time was mentioned before; in general, no state vector exists which would define such a state (Newton and Wigner, 1949). Naturally, this difficulty does not apply to the measurement of stationary states (the possibility of the measurement of the linear momentum was mentioned before). Naturally, the measurements of these quantities also take time, but that does not matter much.

On the whole, it is natural to believe that the "problem of measurement" should be reviewed. This is a fundamental problem, and the present theory underlying its description is surely at least not relativistic. Another difficulty is that the process of observation, which chooses one of the possible outcomes of the measurement in case the original state was a superposition of several possible outcomes, cannot be described by present-day theory. This was mentioned before, in connection with the cubic centimeter tungsten experiment. Let me now come to the last weakness of our present-day theory: the restriction of its area to non-living objects.

. . . .
The Phenomenon of Life

All sciences have a restricted area of interest, but that of physics has grown almost unbelievably in the last four hundred years. Newton's physics was principally interested in the motion of planets but gave a description also to other effects of the gravitational forces, such as that of freely falling bodies here on the earth. But most other phenomena were essentially outside of the interest of Newton's physics. This changed fundamentally with Maxwell's theory of the electric and magnetic forces, which gave also a description of light. But mechanical physics was also greatly extended—sound, for instance, was adequately described.

The next dramatic extension was equally fundamental: it extended the area of physics to microscopic phenomena. As I often mention, the first physics book I read said, "Atoms and molecules may exist, but this is irrelevant from the point of view of physics". But it was natural to catch up with chemistry and to try to give an explanation to properties of bodies, such as density, heat conductivity, specific heat, elastic properties, and others. This led to the development of statistical mechanics, largely in cooperation with the chemists. But the physical chemistry thus developed led to difficulties, most obviously by giving an infinite specific heat to the vacuum, suggesting a very large electromagnetic field therein. This difficulty was resolved by the quantum theory originated by Planck and wonderfully developed later as was implicitly mentioned before.

The question then arises: Are there other phenomena which are still outside of physics' interest? Some deny it, but it seems to me evident that life is not described by present-day physics. For some time I have believed that the mere phenomenon of the observation of the outcomes of measurements shows this. This has been at least superficially eliminated by Zeh's observation and if my own suggestion of a change of the physical theories, mentioned before in connection with the cubic centimeter of tungsten, is accepted. But I am still convinced that life is a fundamental phenomenon entirely outside the present interest of physics, and I hope you agree with me.

There is, of course, a continuous transition between the situation in which the phenomenon I called "life" plays an important role and the phenomenon in the description of which it can be disregarded. The same is true, for instance, with electromagnetic effects—as was demonstrated by Newton for the movement of planets—or for most other effects. It is possible that the phenomenon of life plays virtually no role in the behavior of microbes, and probably of plants; the behavior of these can be, probably, described with a great skill, in terms of the concepts of our physics. But the description of more complex "living beings", and finally of men, I am convinced, is outside the present area of physics.

· · · · · · · · · · ·

Two Questions

We are left with two questions. The first is surely very difficult to answer: Is life the only phenomenon outside the present area of physics? We do not know, but we hope it is the most interesting one. The second question is, naturally: Will we ever be able to make the phenomenon of life part of our science, perhaps still called physics?

We do not know. Man's interest in abstract knowledge, in science, is truly admirable, and we do not know how this interest was born and even less how it will continue. It is not clear that it was needed to secure our survival: man survived about a million years without a significant amount of any knowledge as abstract as is our science. It seems to be a miracle that our sciences were created, and most of them were created in a very short time, perhaps a few thousand years.

Would it be good if our science became complete, if we understood all natural phenomena? Man enjoys working for and expanding his knowledge and understanding. If it were complete, this very dear purpose would disappear. I hope it will not, that it is impossible, even for man, to produce a perfect scientific discipline. But we should strive toward it.

GLOSSARY

· · · · · · · · · · ·

Prepared by Dr. Stacy Kniffen

Anthropic Principle: the idea that humans observe nature and natural phenomena the way they are because if they were otherwise human existence and perception would not be possible.

Benard instability: a non-equilibrium phase transition associated with the formation of periodically spaced convection currents in a fluid which had been initially at rest but subject to a temperature gradient in a gravitational field.

Bohr's Principle of Complementarity: objects may be observed in nature as possessing either particle or wave character but not both simultaneously.

bootstrap theory of nuclear particles: the idea that no nuclear particle is more fundamental than any other but that all must exist together to form a consistent theory.

expectation value: the average value of an observable (defined by the appropriate quantum mechanical operator) for a system described by a particular wavefunction or the method of computing such an average.

Feynman integrals: quantum mechanical calculations of the interaction between particles.

First Law of Thermodynamics: the principle of conservation of energy, stated in the form that heat and mechanical work can be interchanged with each other.

force carrier bosons: particles with integer-valued intrinsic spins (such as photons, W's, and Z's) which are exchanged between two interacting particles. For example, the repulsive electromagnetic force felt between two electrons can be viewed as an exchange of a photon between the two.

Inflationary Universe Model: theory that the initial expansion of the universe was much more rapid than the present rate of expansion.

Lorentz transformation: a mathematical means of describing phenomena observed in a moving reference frame from the point of view of a stationary frame (and vice versa) which is consistent with Einstein's Theory of Special Relativity.

matrix mechanics: another term for quantum mechanics which stresses the representation of operators as matrices and wavefunctions as vectors.

missing mass: hypothesized additional mass distributed throughout the universe which, if it exists, would ensure that the universe will ultimately stop expanding and begin to contract.

no-boundary proposal: a concept of the universe in which space and time are considered finite in size but do not have edges or boundaries. The surface of a sphere is an example of a finite surface without boundaries.

non-equilibrium phase transitions: state changes that occur when a system is away from thermodynamic equilibrium when pressure, temperature, etc. vary with time and spatial position. In such transitions, nonlinear efforts can become important and often lead to complicated behavior.

Oscillating Universe Hypothesis: the theory that the universe repeatedly alternates between periods of expansion from an initial state (as in the Big Bang) and contraction back to such a state (a so-called Big Crunch).

Quantum Gravity/Gravitation: an attempt to combine the concepts of quantum mechanics which describes phenomena on very small scales (e.g., atoms) with the concepts of general relativity which describes the nature of gravity and its effect on massive bodies (e.g., stars, planets).

Quantum Physics/Quantum Mechanics: a branch of physics in which a physical entity (a particle or field) is described not in terms of its exact physical characteristics (e.g., energy) but rather in terms of its probability of being in any of a set of states whose physical characteristics take on certain discrete or quantized values.

quantum wave function: a mathematical quantity which describes a particle or field in terms of probabilities of being in various quantum states rather than physical observables (energy, position, momentum). These observables are associated with various mathematical operations on the wave function and are called quantum mechanical operators.

Pauli exclusion principle: no two fermions can be in exactly the same quantum state. In an atom, this means that no two electrons (fermions) can have the same energy, orbital angular momentum, and spin angular momentum.

Phase transition: the change in state of matter as from a liquid to a solid (freezing) or a liquid to a gas (boiling). Another example would be the transition from the normal to the superfluid state in liquid helium.

Principle of Explanation: a metascientific principle that affirms the existence of an explanation for everything that exists because it postulates rationality at all levels of reality.

Regge pole theory: a theory of the scattering of elementary particles in which these particles lie on "Regge trajectories" that relate their masses and spins.

Second Law of Thermodynamics: the law that closed systems always proceed from order to disorder; entropy (a measure of disorder) never decreases.

singularity/-ies: a point in which all the universe is compressed into an infinitely dense mass. In mathematics, a point where a function is infinite in value.

S-matrix: quantum mechanical operator associated with intrinsic or spin angular momentum. Particles may possess either integer or half-integer units of spin.

Superstrings: theoretical constructs that seek to explain all observed particles and forces, including gravitation, in terms of the excitations of fundamental one-dimensional objects called strings.

symmetry group: a set of operations that do not change the quantity they operate on. For example, rotation and reflection of a circle produces another circle.

uncertainty principle: the quantum mechanical limitation on the accuracy of certain pairs of physical measurements. For example, both the position and momentum of a particle cannot both be known simultaneously because, according to quantum mechanics, the observation of one quantity affects the other.

unified field theory: a theory that combines two or more of the four theories of natural forces (electromagnetic, weak nuclear, strong nuclear, or gravity) into a single consistent theory.

Vacuum Fluctuation Model: the quantum mechanical prediction that (electromagnetic) fields exist in vacuum apart from sources of the fields (i.e., charge and current) that are necessary in the classical field theory.

wave mechanics: a term for quantum mechanics which stresses the fact that particles may have wave characteristics.

Z particle: an uncharged boson (integer spin) which is partly responsible for carrying the weak nuclear force. The Z is similar to the photon but much more massive.

zerospace (pointlike) fermions: particles with half-integer intrinsic spins which have mass but no measurable volume. Fermions obey the Pauli exclusion principle.

INDEX

· · · · · · · · · ·

Lightning Source UK Ltd.
Milton Keynes UK
14 March 2011

169258UK00001B/66/A